KT-477-585

CALCULUS DEMYSTIFIED

Other Titles in the McGraw-Hill Demystified Series

Algebra Demystified by Rhonda Huettenmueller
Astronomy Demystified by Stan Gibilisco
Physics Demystified by Stan Gibilisco

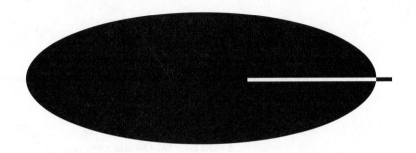

CALCULUS DEMYSTIFIED

STEVEN G. KRANTZ

McGRAW-HILL
New York   Chicago   San Francisco   Lisbon   London   Madrid
Mexico City   Milan   New Delhi   San Juan   Seoul
Singapore   Sydney   Toronto

Cataloging-in-Publication Data is on file with the Library of Congress.

McGraw-Hill

A Division of The McGraw·Hill Companies

Copyright © 2003 by The McGraw-Hill Companies, Inc. All rights reserved. Printed in the United States of America. Except as permitted under the United States Copyright Act of 1976, no part of this publication may be reproduced or distributed in any form or by any means, or stored in a data base or retrieval system, without the prior written permission of the publisher.

13 14   DOC/DOC   0 9 8 7 6

ISBN 0-07-139308-0

The sponsoring editor for this book was Scott Grillo and the production supervisor was Sherri Souffrance. It was set in Times Roman by Keyword Publishing Services.

Printed and bound by RR Donnelley

This book was printed on recycled, acid-free paper containing a minimum of 50% recycled, de-inked fiber.

McGraw-Hill books are available at special quantity discounts to use as premiums and sales promotions, or for use in corporate training programs. For more information, please write to the Director of Special Sales, Professional Publishing, McGraw-Hill, Two Penn Plaza, New York, NY 10121-2298. Or contact your local bookstore.

Information contained in this work has been obtained by The McGraw-Hill Companies, Inc. ("McGraw-Hill") from sources believed to be reliable. However, neither McGraw-Hill nor its authors guarantee the accuracy or completeness of any information published herein, and neither McGraw-Hill nor its authors shall be responsible for any errors, omissions, or damages arising out of use of this information. This work is published with the understanding that McGraw-Hill and its authors are supplying information but are not attempting to render engineering or other professional services. If such services are required, the assistance of an appropriate professional should be sought.

To Archimedes, Pierre de Fermat, Isaac Newton, and Gottfried Wilhelm von Leibniz, the fathers of calculus

CONTENTS

Contents

Contents

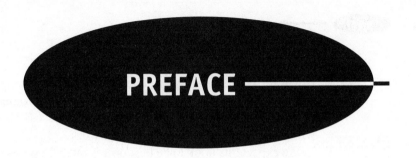

PREFACE

Calculus is one of the milestones of Western thought. Building on ideas of Archimedes, Fermat, Newton, Leibniz, Cauchy, and many others, the calculus is arguably the cornerstone of modern science. Any well-educated person should at least be acquainted with the ideas of calculus, and a scientifically literate person must know calculus solidly.

Calculus has two main aspects: differential calculus and integral calculus. Differential calculus concerns itself with rates of change. Various types of change, both mathematical and physical, are described by a mathematical quantity called the *derivative*. Integral calculus is concerned with a generalized type of addition, or amalgamation, of quantities. Many kinds of summation, both mathematical and physical, are described by a mathematical quantity called the *integral*.

What makes the subject of calculus truly powerful and seminal is the Fundamental Theorem of Calculus, which shows how an integral may be calculated by using the theory of the derivative. The Fundamental Theorem enables a number of important conceptual breakthroughs and calculational techniques. It makes the subject of differential equations possible (in the sense that it gives us ways to *solve* these equations).

Calculus Demystified explains this panorama of ideas in a step-by-step and accessible manner. The author, a renowned teacher and expositor, has a strong sense of the level of the students who will read this book, their backgrounds and their strengths, and can present the material in accessible morsels that the student can study on his own. Well-chosen examples and cognate exercises will reinforce the ideas being presented. Frequent review, assessment, and application of the ideas will help students to retain and to internalize all the important concepts of calculus.

We envision a book that will give the student a firm grounding in calculus. The student who has mastered this book will be able to go on to study physics, engineering, chemistry, computational biology, computer science, and other basic scientific areas that use calculus.

Calculus Demystified will be a valuable addition to the self-help literature. Written by an accomplished and experienced teacher (the author of *How to Teach Mathematics*), this book will aid the student who is working without a teacher.

It will provide encouragement and reinforcement as needed, and diagnostic exercises will help the student to measure his or her progress. A comprehensive exam at the end of the book will help the student to assess his mastery of the subject, and will point to areas that require further work.

We expect this book to be the cornerstone of a series of elementary mathematics books of the same tenor and utility.

STEVEN G. KRANTZ
St. Louis, Missouri

CHAPTER 1

Basics

1.0 Introductory Remarks

Calculus is one of the most important parts of mathematics. It is fundamental to all of modern science. How could one part of mathematics be of such central importance? It is because calculus gives us the tools to study *rates of change* and *motion*. All analytical subjects, from biology to physics to chemistry to engineering to mathematics, involve studying quantities that are growing or shrinking or moving—in other words, they are *changing*. Astronomers study the motions of the planets, chemists study the interaction of substances, physicists study the interactions of physical objects. All of these involve change and motion.

   In order to study calculus effectively, you must be familiar with cartesian geometry, with trigonometry, and with functions. We will spend this first chapter reviewing the essential ideas. Some readers will study this chapter selectively, merely reviewing selected sections. Others will, for completeness, wish to review all the material. The main point is to get started on calculus (Chapter 2).

1.1 Number Systems

The number systems that we use in calculus are the *natural numbers*, the *integers*, the *rational numbers*, and the *real numbers*. Let us describe each of these:

- The natural numbers are the system of positive counting numbers 1, 2, 3, ...
  We denote the set of all natural numbers by $\mathbb{N}$.

- The integers are the positive and negative whole numbers and zero: $\dots, -3, -2, -1, 0, 1, 2, 3, \dots$. We denote the set of all integers by $\mathbb{Z}$.
- The rational numbers are quotients of integers. Any number of the form p/q, with $p, q \in \mathbb{Z}$ and $q \neq 0$, is a rational number. We say that p/q and r/s represent the *same rational number* precisely when $ps = qr$. Of course you know that in displayed mathematics we write fractions in this way:

$$\frac{1}{2} + \frac{2}{3} = \frac{7}{6}.$$

- The real numbers are the set of all decimals, both terminating and non-terminating. This set is rather sophisticated, and bears a little discussion. A decimal number of the form

$$x = 3.16792$$

is actually a rational number, for it represents

$$x = 3.16792 = \frac{316792}{100000}.$$

A decimal number of the form

$$m = 4.27519191919\dots,$$

with a group of digits that repeats itself interminably, is also a rational number. To see this, notice that

$$100 \cdot m = 427.519191919\dots$$

and therefore we may subtract:

$$100m = 427.519191919\dots$$
$$m = 4.275191919\dots$$

Subtracting, we see that

$$99m = 423.244$$

or

$$m = \frac{423244}{99000}.$$

So, as we asserted, m is a rational number or quotient of integers.

The third kind of decimal number is one which has a non-terminating decimal expansion *that does not keep repeating*. An example is $3.14159265\dots$. This is the decimal expansion for the number that we ordinarily call π. Such a number is *irrational*, that is, it *cannot* be expressed as the quotient of two integers.

In summary: There are three types of real numbers: (i) terminating decimals, (ii) non-terminating decimals that repeat, (iii) non-terminating decimals that do not repeat. Types (i) and (ii) are rational numbers. Type (iii) are irrational numbers.

You Try It: What type of real number is 3.41287548754875 . . . ? Can you express this number in more compact form?

1.2   Coordinates in One Dimension

We envision the real numbers as laid out on a line, and we locate real numbers from left to right on this line. If $a < b$ are real numbers then a will lie to the left of b on this line. See Fig. 1.1.

Fig. 1.1

EXAMPLE 1.1
On a real number line, plot the numbers −4, −1, 2, 6. Also plot the sets $S = \{x \in \mathbb{R}: -8 \leq x < -5\}$ and $T = \{t \in \mathbb{R}: 7 < t \leq 9\}$. Label the plots.

SOLUTION
Figure 1.2 exhibits the indicated points and the two sets. These sets are called *half-open intervals* because each set includes one endpoint and not the other.

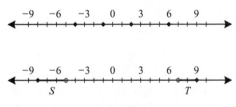

Fig. 1.2

Math Note: The notation $S = \{x \in \mathbb{R}: -8 \leq x < -5\}$ is called *set builder notation*. It says that S is the set of all numbers x such that x is greater than or equal to −8 and less than 5. We will use set builder notation throughout the book.

If an interval contains both its endpoints, then it is called a *closed interval*. If an interval omits both its endpoints, then it is called an *open interval*. See Fig. 1.3.

closed interval   open interval

Fig. 1.3

EXAMPLE 1.2

Find the set of points that satisfy $x - 2 < 4$ and exhibit it on a number line.

SOLUTION

We solve the inequality to obtain $x < 6$. The set of points satisfying this inequality is exhibited in Fig. 1.4.

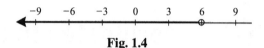

Fig. 1.4

EXAMPLE 1.3

Find the set of points that satisfy the condition

$$|x + 3| \leq 2 \qquad (*)$$

and exhibit it on a number line.

SOLUTION

In case $x + 3 \geq 0$ then $|x + 3| = x + 3$ and we may write condition $(*)$ as

$$x + 3 \leq 2$$

or

$$x \leq -1.$$

Combining $x + 3 \geq 0$ and $x \leq -1$ gives $-3 \leq x \leq -1$.

On the other hand, if $x + 3 < 0$ then $|x + 3| = -(x + 3)$. We may then write condition $(*)$ as

$$-(x + 3) \leq 2$$

or

$$-5 \leq x.$$

Combining $x + 3 < 0$ and $-5 \leq x$ gives $-5 \leq x < -3$.

We have found that our inequality $|x + 3| \leq 2$ is true precisely when either $-3 \leq x \leq -1$ or $-5 \leq x < -3$. Putting these together yields $-5 \leq x \leq -1$. We display this set in Fig. 1.5.

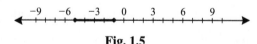

Fig. 1.5

You Try It: Solve the inequality $|x - 4| > 1$. Exhibit your answer on a number line.

You Try It: On a real number line, sketch the set $\{x : x^2 - 1 < 3\}$.

1.3   Coordinates in Two Dimensions

We locate points in the plane by using two coordinate lines (instead of the single line that we used in one dimension). Refer to Fig. 1.6. We determine the coordinates of the given point P by first determining the x-displacement, or (signed) distance from the y-axis and then determining the y-displacement, or (signed) distance from the x-axis. We refer to this coordinate system as (x, y)-coordinates or *Cartesian coordinates*. The idea is best understood by way of some examples.

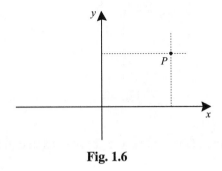

Fig. 1.6

EXAMPLE 1.4

Plot the points $P = (3, -2)$, $Q = (-4, 6)$, $R = (2, 5)$, $S = (-5, -3)$.

SOLUTION

The first coordinate 3 of the point P tells us that the point is located 3 units to the *right* of the y-axis (because 3 is *positive*). The second coordinate -2 of the point P tells us that the point is located 2 units *below* the x-axis (because -2 is negative). See Fig. 1.7.

The first coordinate -4 of the point Q tells us that the point is located 4 units to the *left* of the y-axis (because -4 is *negative*). The second coordinate 6 of the point Q tells us that the point is located 6 units *above* the x-axis (because 6 is positive). See Fig. 1.7.

The first coordinate 2 of the point R tells us that the point is located 2 units to the *right* of the y-axis (because 2 is *positive*). The second coordinate 5 of the point R tells us that the point is located 5 units *above* the x-axis (because 5 is positive). See Fig. 1.7.

The first coordinate -5 of the point S tells us that the point is located 5 units to the *left* of the y-axis (because -5 is *negative*). The second coordinate -3 of the point S tells us that the point is located 3 units *below* the x-axis (because -3 is negative). See Fig. 1.7.

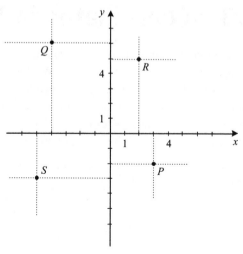

Fig. 1.7

EXAMPLE 1.5
Give the coordinates of the points *X, Y, Z, W* exhibited in Fig. 1.8.

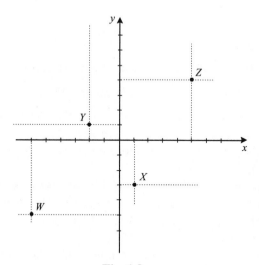

Fig. 1.8

SOLUTION
The point X is 1 unit to the right of the y-axis and 3 units below the x-axis. Therefore its coordinates are $(1, -3)$.

The point Y is 2 units to the left of the y-axis and 1 unit above the x-axis. Therefore its coordinates are $(-2, 1)$.

The point Z is 5 units to the right of the y-axis and 4 units above the x-axis. Therefore its coordinates are $(5, 4)$.

The point W is 6 units to the left of the y-axis and 5 units below the x-axis. Therefore its coordinates are $(-6, -5)$.

You Try It: Sketch the points $(3, -5)$, $(2, 4)$, $(\pi, \pi/3)$ on a set of axes. Sketch the set $\{(x, y): x = 3\}$ on another set of axes.

EXAMPLE 1.6

Sketch the set of points $\ell = \{(x, y): y = 3\}$. Sketch the set of points $k = \{(x, y): x = -4\}$.

SOLUTION

The set ℓ consists of all points with y-coordinate equal to 3. This is the set of all points that lie 3 units above the x-axis. We exhibit ℓ in Fig. 1.9. It is a horizontal line.

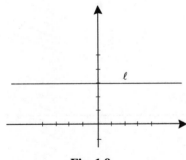

Fig. 1.9

The set k consists of all points with x-coordinate equal to -4. This is the set of all points that lie 4 units to the left of the y-axis. We exhibit k in Fig. 1.10. It is a vertical line.

EXAMPLE 1.7

Sketch the set of points $S = \{(x, y): x > 2\}$ on a pair of coordinate axes.

SOLUTION

Notice that the set S contains all points with x-coordinate greater than 2. These will be all points to the right of the vertical line $x = 2$. That set is exhibited in Fig. 1.11.

You Try It: Sketch the set $\{(x, y): x + y < 4\}$.

You Try It: Identify the set (using set builder notation) that is shown in Fig. 1.12.

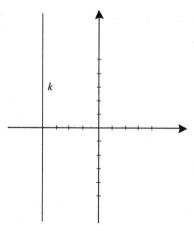

Fig. 1.10

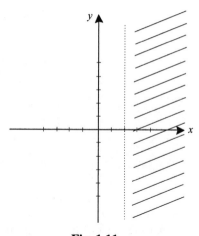

Fig. 1.11

1.4   The Slope of a Line in the Plane

A line in the plane may rise gradually from left to right, or it may rise quite steeply from left to right (Fig. 1.13). Likewise, it could fall gradually from left to right, or it could fall quite steeply from left to right (Fig. 1.14). The number "slope" differentiates among these different rates of rise or fall.

Look at Fig. 1.15. We use the two points $P = (p_1, p_2)$ and $Q = (q_1, q_2)$ to calculate the slope. It is

$$m = \frac{q_2 - p_2}{q_1 - p_1}.$$

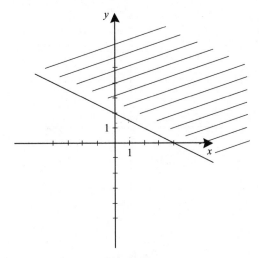

Fig. 1.12

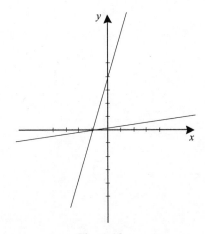

Fig. 1.13

It turns out that, no matter which two points we may choose on a given line, this calculation will always give the same answer for slope.

EXAMPLE 1.8

Calculate the slope of the line in Fig. 1.16.

SOLUTION

We use the points $P = (-1, 0)$ and $Q = (1, 3)$ to calculate the slope of this line:

$$m = \frac{3 - 0}{1 - (-1)} = \frac{3}{2}.$$

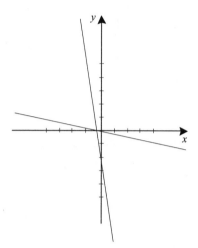

Fig. 1.14

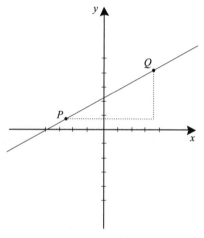

Fig. 1.15

We could just as easily have used the points $P = (-1, 0)$ and $R = (3, 6)$ to calculate the slope:

$$m = \frac{6 - 0}{3 - (-1)} = \frac{6}{4} = \frac{3}{2}.$$

If a line has slope m, then, for each unit of motion from left to right, the line rises m units. In the last example, the line rises 3/2 units for each unit of motion to the right. Or one could say that the line rises 3 units for each 2 units of motion to the right.

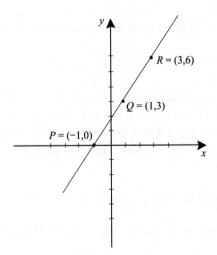

Fig. 1.16

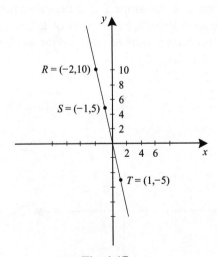

Fig. 1.17

EXAMPLE 1.9
Calculate the slope of the line in Fig. 1.17.

SOLUTION
We use the points $R = (-2, 10)$ and $T = (1, -5)$ to calculate the slope of this line:

$$m = \frac{10 - (-5)}{(-2) - 1} = -5.$$

We could just as easily have used the points $S = (-1, 5)$ and $T = (1, -5)$:

$$m = \frac{5 - (-5)}{-1 - 1} = -5.$$

In this example, the line falls 5 units for each 1 unit of left-to-right motion. The negativity of the slope indicates that the line is falling.

The concept of slope is undefined for a vertical line. Such a line will have any two points with the same x-coordinate, and calculation of slope would result in division by 0.

You Try It: What is the slope of the line $y = 2x + 8$?

You Try It: What is the slope of the line $y = 5$? What is the slope of the line $x = 3$?

Two lines are perpendicular precisely when their slopes are negative reciprocals. This makes sense: If one line has slope 5 and the other has slope $-1/5$ then we see that the first line rises 5 units for each unit of left-to-right motion while the second line falls 1 unit for each 5 units of left-to-right motion. So the lines must be perpendicular. See Fig. 1.18(a).

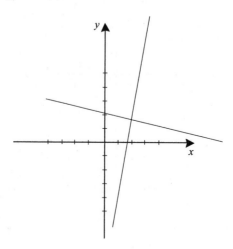

Fig. 1.18(a)

You Try It: Sketch the line that is perpendicular to $x + 2y = 7$ and passes through $(1, 4)$.

Note also that two lines are parallel precisely when they have the *same slope*. See Fig. 1.18(b).

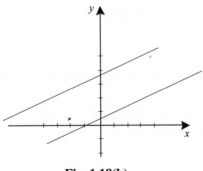

Fig. 1.18(b)

1.5   The Equation of a Line

The equation of a line in the plane will describe—in compact form—all the points that lie on that line. We determine the equation of a given line by writing its slope in two different ways and then equating them. Some examples best illustrate the idea.

EXAMPLE 1.10

Determine the equation of the line with slope 3 that passes through the point $(2, 1)$.

SOLUTION

Let (x, y) be a variable point on the line. Then we can use that variable point together with $(2, 1)$ to calculate the slope:

$$m = \frac{y - 1}{x - 2}.$$

On the other hand, we are given that the slope is $m = 3$. We may equate the two expressions for slope to obtain

$$3 = \frac{y - 1}{x - 2}. \qquad (*)$$

This may be simplified to $y = 3x - 5$.

Math Note: The form $y = 3x - 5$ for the equation of a line is called the *slope-intercept form*. The slope is 3 and the line passes through $(0, 5)$ (its y-intercept).

Math Note: Equation $(*)$ may be rewritten as $y - 1 = 3(x - 2)$. In general, the line with slope m that passes through the point (x_0, y_0) can be written as $y - y_0 = m(x - x_0)$. This is called the *point-slope* form of the equation of a line.

You Try It: Write the equation of the line that passes through the point $(-3, 2)$ and has slope 4.

EXAMPLE 1.11

Write the equation of the line passing through the points $(-4, 5)$ and $(6, 2)$.

SOLUTION

Let (x, y) be a variable point on the line. Using the points (x, y) and $(-4, 5)$, we may calculate the slope to be

$$m = \frac{y - 5}{x - (-4)}.$$

On the other hand, we may use the points $(-4, 5)$ and $(6, 2)$ to calculate the slope:

$$m = \frac{2 - 5}{6 - (-4)} = \frac{-3}{10}.$$

Equating the two expressions for slope, we find that

$$\frac{y - 5}{x + 4} = \frac{-3}{10}.$$

Simplifying this identity, we find that the equation of our line is

$$y - 5 = \frac{-3}{10} \cdot (x + 4).$$

You Try It: Find the equation of the line that passes through the points $(2, -5)$ and $(-6, 1)$.

In general, the line that passes through points (x_0, y_0) and (x_1, y_1) has equation

$$\frac{y - y_0}{x - x_0} = \frac{y_1 - y_0}{x_1 - x_0}.$$

This is called the *two-point form* of the equation of a line.

EXAMPLE 1.12

Find the line perpendicular to $y = 3x - 6$ that passes through the point $(5, 4)$.

SOLUTION

We know from the Math Note immediately after Example 1.10 that the given line has slope 3. Thus the line we seek (the perpendicular line) has slope $-1/3$. Using the point-slope form of a line, we may immediately write the equation of the line with slope $-1/3$ and passing through $(5, 4)$ as

$$y - 4 = \frac{-1}{3} \cdot (x - 5).$$

Summary: We determine the equation of a line in the plane by finding two expressions for the slope and equating them.

If a line has slope m and passes through the point (x_0, y_0) then it has equation

$$y - y_0 = m(x - x_0).$$

This is the point-slope form of a line.

If a line passes through the points (x_0, y_0) and (x_1, y_1) then it has equation

$$\frac{y - y_0}{x - x_0} = \frac{y_1 - y_0}{x_1 - x_0}.$$

This is the two-point form of a line.

You Try It: Find the line perpendicular to $2x + 5y = 10$ that passes through the point $(1, 1)$. Now find the line that is parallel to the given line and passes through $(1, 1)$.

1.6 Loci in the Plane

The most interesting sets of points to graph are collections of points that are defined by an equation. We call such a graph the *locus* of the equation. We cannot give all the theory of loci here, but instead consider a few examples. See [SCH2] for more on this matter.

EXAMPLE 1.13

Sketch the graph of $\{(x, y): y = x^2\}$.

SOLUTION

It is convenient to make a table of values:

x	$y = x^2$
-3	9
-2	4
-1	1
0	0
1	1
2	4
3	9

We plot these points on a single set of axes (Fig. 1.19). Supposing that the curve we seek to draw is a smooth interpolation of these points (calculus will later show us that this supposition is correct), we find that our curve is as shown in Fig. 1.20. This curve is called a *parabola*.

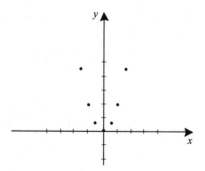

Fig. 1.19

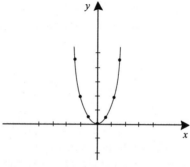

Fig. 1.20

EXAMPLE 1.14

Sketch the graph of the curve $\{(x, y): y = x^3\}$.

SOLUTION

It is convenient to make a table of values:

x	$y = x^3$
-3	-27
-2	-8
-1	-1
0	0
1	1
2	8
3	27

We plot these points on a single set of axes (Fig. 1.21). Supposing that the curve we seek to draw is a smooth interpolation of these points (calculus will later show us that this supposition is correct), we find that our curve is as shown in Fig. 1.22. This curve is called a *cubic*.

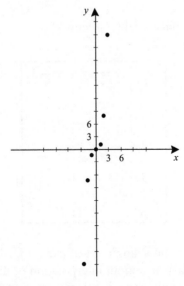

Fig. 1.21

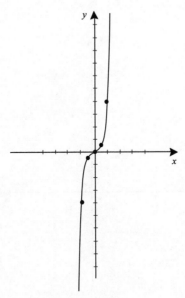

Fig. 1.22

You Try It: Sketch the graph of the locus $|x| = |y|$.

EXAMPLE 1.15
Sketch the graph of the curve $y = x^2 + x - 1$.

SOLUTION

It is convenient to make a table of values:

x	$y = x^2 + x - 1$
-4	11
-3	5
-2	1
-1	-1
0	-1
1	1
2	5
3	11

We plot these points on a single set of axes (Fig. 1.23). Supposing that the curve we seek to draw is a smooth interpolation of these points (calculus will later show us that this supposition is correct), we find that our curve is as shown in Fig. 1.24. This is another example of a parabola.

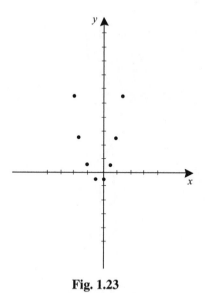

Fig. 1.23

You Try It: Sketch the locus $y^2 = x^3 + x + 1$ on a set of axes.

The reader unfamiliar with cartesian geometry and the theory of loci would do well to consult [SCH2].

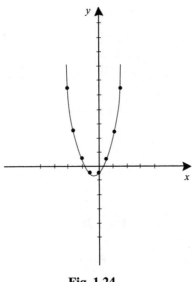

Fig. 1.24

1.7 Trigonometry

Here we give a whirlwind review of basic ideas of trigonometry. The reader who needs a more extensive review should consult [SCH1].

When we first learn trigonometry, we do so by studying right triangles and measuring angles in degrees. Look at Fig. 1.25. In calculus, however, it is convenient to study trigonometry in a more general setting, and to measure angles in radians.

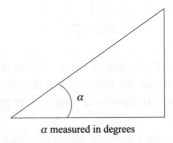

α measured in degrees

Fig. 1.25

Angles will be measured by rotation along the unit circle in the plane, beginning at the positive x-axis. See Fig. 1.26. Counterclockwise rotation corresponds to positive angles, and clockwise rotation corresponds to negative angles. Refer to Fig. 1.27. The *radian measure* of an angle is defined to be the length of the arc of the unit circle that the angle subtends with the positive x-axis (together with an appropriate $+$ or $-$ sign).

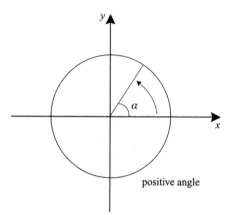

Fig. 1.26

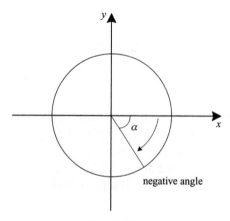

Fig. 1.27

In degree measure, one full rotation about the unit circle is 360°; in radian measure, one full rotation about the circle is just the circumference of the circle or 2π. Let us use the symbol θ to denote an angle. The principle of proportionality now tells us that

$$\frac{\text{degree measure of } \theta}{360°} = \frac{\text{radian measure of } \theta}{2\pi}.$$

In other words

$$\text{radian measure of } \theta = \frac{\pi}{180} \cdot (\text{degree measure of } \theta)$$

and

$$\text{degree measure of } \theta = \frac{180}{\pi} \cdot (\text{radian measure of } \theta).$$

EXAMPLE 1.16

Sketch the angle with radian measure $\pi/6$. Give its equivalent degree measure.

SOLUTION

  Since

$$\frac{\pi/6}{2\pi} = \frac{1}{12},$$

the angle subtends an arc of the unit circle corresponding to 1/12 of the full circumference. Since $\pi/6 > 0$, the angle represents a counterclockwise rotation. It is illustrated in Fig. 1.28.

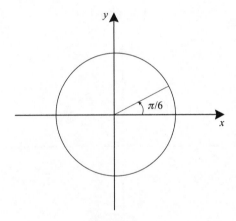

Fig. 1.28

  The degree measure of this angle is

$$\frac{180}{\pi} \cdot \frac{\pi}{6} = 30°.$$

Math Note: In this book we *always* use radian measure for angles. (The reason is that it makes the formulas of calculus turn out to be simpler.) Thus, for example, if we refer to "the angle $2\pi/3$" then it should be understood that this is an angle in radian measure. See Fig. 1.29.

  Likewise, if we refer to the angle 3 it is also understood to be radian measure. We sketch this last angle by noting that 3 is approximately 0.477 of a full rotation 2π—refer to Fig. 1.30.

You Try It: Sketch the angles -2, 1, π, $3\pi/2$, 10—all on the same coordinate figure. Of course use radian measure.

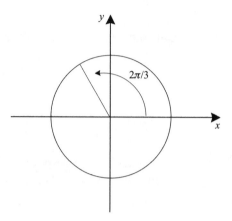

Fig. 1.29

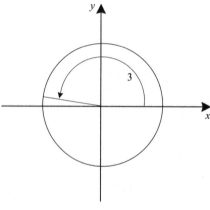

Fig. 1.30

EXAMPLE 1.17

Several angles are sketched in Fig. 1.31, and both their radian and degree measures given.

If θ is an angle, let (x, y) be the coordinates of the terminal point of the corresponding radius (called the *terminal radius*) on the unit circle. We call $P = (x, y)$ the *terminal point* corresponding to θ. Look at Fig. 1.32. The number y is called the *sine* of θ and is written $\sin \theta$. The number x is called the *cosine* of θ and is written $\cos \theta$.

Since $(\cos \theta, \sin \theta)$ are coordinates of a point on the unit circle, the following two fundamental properties are immediate:

(1)   For any number θ,

$$(\sin \theta)^2 + (\cos \theta)^2 = 1.$$

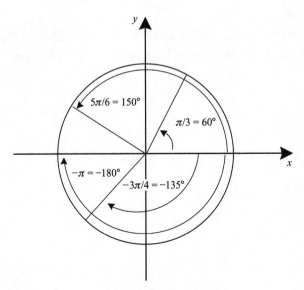

Fig. 1.31

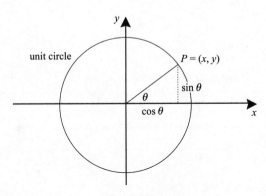

Fig. 1.32

(2)   For any number θ,

$$-1 \leq \cos \theta \leq 1 \qquad \text{and} \qquad -1 \leq \sin \theta \leq 1.$$

Math Note: It is common to write $\sin^2 \theta$ to mean $(\sin \theta)^2$ and $\cos^2 \theta$ to mean $(\cos \theta)^2$.

EXAMPLE 1.18

Compute the sine and cosine of $\pi/3$.

SOLUTION

We sketch the terminal radius and associated triangle (see Fig. 1.33). This is a 30–60–90 triangle whose sides have ratios $1 : \sqrt{3} : 2$. Thus

$$\frac{1}{x} = 2 \qquad \text{or} \qquad x = \frac{1}{2}.$$

Likewise,

$$\frac{y}{x} = \sqrt{3} \qquad \text{or} \qquad y = \sqrt{3}x = \frac{\sqrt{3}}{2}.$$

It follows that

$$\sin \frac{\pi}{3} = \frac{\sqrt{3}}{2}$$

and

$$\cos \frac{\pi}{3} = \frac{1}{2}.$$

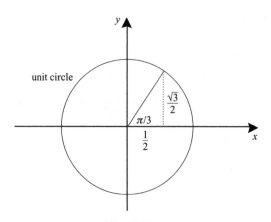

Fig. 1.33

You Try It: The cosine of a certain angle is $2/3$. The angle lies in the fourth quadrant. What is the sine of the angle?

Math Note: Notice that if θ is an angle then θ and $\theta + 2\pi$ have the same terminal radius and the same terminal point (for adding 2π just adds one more trip around the circle—look at Fig. 1.34).

As a result,

$$\sin \theta = x = \sin(\theta + 2\pi)$$

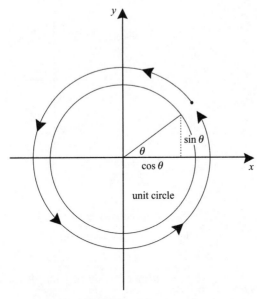

Fig. 1.34

and

$$\cos\theta = y = \cos(\theta + 2\pi).$$

We say that the sine and cosine functions have *period* 2π: the functions repeat themselves every 2π units.

In practice, when we calculate the trigonometric functions of an angle θ, we reduce it by multiples of 2π so that we can consider an equivalent angle θ', called the *associated principal angle,* satisfying $0 \le \theta' < 2\pi$. For instance,

$$15\pi/2 \quad \text{has associated principal angle}$$
$$3\pi/2 \quad (\text{since } 15\pi/2 - 3\pi/2 = 3 \cdot 2\pi)$$

and

$$-10\pi/3 \quad \text{has associated principal angle}$$
$$2\pi/3 \quad (\text{since } -10\pi/3 - 2\pi/3 = -12\pi/3 = -2 \cdot 2\pi).$$

You Try It: What are the principal angles associated with 7π, $11\pi/2$, $8\pi/3$, $-14\pi/5$, $-16\pi/7$?

What does the concept of angle and sine and cosine that we have presented here have to do with the classical notion using triangles? Notice that any angle θ such that $0 \le \theta < \pi/2$ has associated to it a right triangle in the first quadrant, with vertex on the unit circle, such that the base is the segment connecting $(0, 0)$ to $(x, 0)$ and the height is the segment connecting $(x, 0)$ to (x, y). See Fig. 1.35.

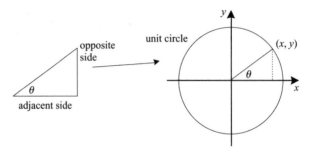

Fig. 1.35

Then

$$\sin \theta = y = \frac{y}{1} = \frac{\text{opposite side of triangle}}{\text{hypotenuse}}$$

and

$$\cos \theta = x = \frac{x}{1} = \frac{\text{adjacent side of triangle}}{\text{hypotenuse}}.$$

Thus, for angles θ between 0 and $\pi/2$, the new definition of sine and cosine using the unit circle is clearly equivalent to the classical definition using adjacent and opposite sides and the hypotenuse. For other angles θ, the classical approach is to reduce to this special case by subtracting multiples of $\pi/2$. Our approach using the unit circle is considerably clearer because it makes the signatures of sine and cosine obvious.

Besides sine and cosine, there are four other trigonometric functions:

$$\tan \theta = \frac{y}{x} = \frac{\sin \theta}{\cos \theta}$$

$$\cot \theta = \frac{x}{y} = \frac{\cos \theta}{\sin \theta}$$

$$\sec \theta = \frac{1}{x} = \frac{1}{\cos \theta}$$

$$\csc \theta = \frac{1}{y} = \frac{1}{\sin \theta}.$$

Whereas sine and cosine have domain the entire real line, we notice that $\tan \theta$ and $\sec \theta$ are undefined at odd multiples of $\pi/2$ (because cosine will vanish there) and $\cot \theta$ and $\csc \theta$ are undefined at even multiples of $\pi/2$ (because sine will vanish there). The graphs of the six trigonometric functions are shown in Fig. 1.36.

EXAMPLE 1.19

Compute all the trigonometric functions for the angle $\theta = 11\pi/4$.

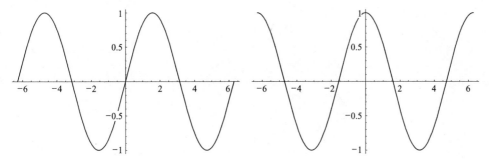

Fig. 1.36(a)   Graphs of $y = \sin x$ and $y = \cos x$.

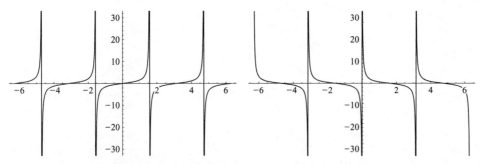

Fig. 1.36(b)   Graphs of $y = \tan x$ and $y = \cot x$.

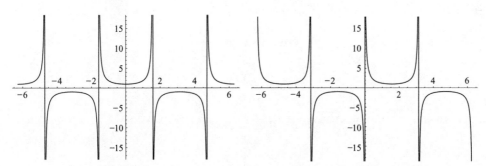

Fig. 1.36(c)   Graphs of $y = \sec x$ and $y = \csc x$.

SOLUTION

   We first notice that the principal associated angle is $3\pi/4$, so we deal with that angle. Figure 1.37 shows that the triangle associated to this angle is an isosceles right triangle with hypotenuse 1.

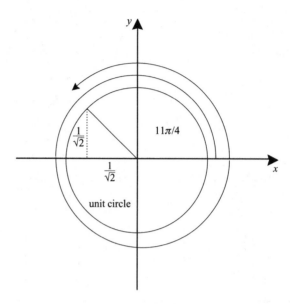

Fig. 1.37

Therefore $x = -1/\sqrt{2}$ and $y = 1/\sqrt{2}$. It follows that

$$\sin \theta = y = \frac{1}{\sqrt{2}}$$

$$\cos \theta = x = -\frac{1}{\sqrt{2}}$$

$$\tan \theta = \frac{y}{x} = -1$$

$$\cot \theta = \frac{x}{y} = -1$$

$$\sec \theta = \frac{1}{x} = -\sqrt{2}$$

$$\csc \theta = \frac{1}{y} = \sqrt{2}.$$

Similar calculations allow us to complete the following table for the values of the trigonometric functions at the principal angles which are multiples of $\pi/6$ or $\pi/4$.

Angle	Sin	Cos	Tan	Cot	Sec	Csc
0	0	1	0	undef	1	undef
$\pi/6$	$1/2$	$\sqrt{3}/2$	$1/\sqrt{3}$	$\sqrt{3}$	$2/\sqrt{3}$	2
$\pi/4$	$\sqrt{2}/2$	$\sqrt{2}/2$	1	1	$\sqrt{2}$	$\sqrt{2}$
$\pi/3$	$\sqrt{3}/2$	$1/2$	$\sqrt{3}$	$1/\sqrt{3}$	2	$2/\sqrt{3}$
$\pi/2$	1	0	undef	0	undef	1
$2\pi/3$	$\sqrt{3}/2$	$-1/2$	$-\sqrt{3}$	$-1/\sqrt{3}$	-2	$2/\sqrt{3}$
$3\pi/4$	$\sqrt{2}/2$	$-\sqrt{2}/2$	-1	-1	$-\sqrt{2}$	$\sqrt{2}$
$5\pi/6$	$1/2$	$-\sqrt{3}/2$	$-1/\sqrt{3}$	$-\sqrt{3}$	$-2/\sqrt{3}$	2
π	0	-1	0	undef	-1	undef
$7\pi/6$	$-1/2$	$-\sqrt{3}/2$	$1/\sqrt{3}$	$\sqrt{3}$	$-2/\sqrt{3}$	-2
$5\pi/4$	$-\sqrt{2}/2$	$-\sqrt{2}/2$	1	1	$-\sqrt{2}$	$-\sqrt{2}$
$4\pi/3$	$-\sqrt{3}/2$	$-1/2$	$\sqrt{3}$	$1/\sqrt{3}$	-2	$-2/\sqrt{3}$
$3\pi/2$	-1	0	undef	0	undef	-1
$5\pi/3$	$-\sqrt{3}/2$	$1/2$	$-\sqrt{3}$	$-1/\sqrt{3}$	2	$-2/\sqrt{3}$
$7\pi/4$	$-\sqrt{2}/2$	$\sqrt{2}/2$	-1	-1	$\sqrt{2}$	$-\sqrt{2}$
$11\pi/6$	$-1/2$	$\sqrt{3}/2$	$-1/\sqrt{3}$	$-\sqrt{3}$	$2/\sqrt{3}$	-2

Besides properties (1) and (2) above, there are certain identities which are fundamental to our study of the trigonometric functions. Here are the principal ones:

(3) $\tan^2\theta + 1 = \sec^2\theta$

(4) $\cot^2\theta + 1 = \csc^2\theta$

(5) $\sin(\theta + \psi) = \sin\theta\cos\psi + \cos\theta\sin\psi$

(6) $\cos(\theta + \psi) = \cos\theta\cos\psi - \sin\theta\sin\psi$

(7) $\sin(2\theta) = 2\sin\theta\cos\theta$

(8) $\cos(2\theta) = \cos^2\theta - \sin^2\theta$

(9) $\sin(-\theta) = -\sin\theta$

(10) $\cos(-\theta) = \cos\theta$

(11) $\sin^2\theta = \dfrac{1 - \cos 2\theta}{2}$

(12) $\cos^2\theta = \dfrac{1 + \cos 2\theta}{2}$

EXAMPLE 1.20

Prove identity number (3).

SOLUTION

We have

$$
\begin{aligned}
\tan^2 \theta + 1 &= \frac{\sin^2 \theta}{\cos^2 \theta} + 1 \\
&= \frac{\sin^2 \theta}{\cos^2 \theta} + \frac{\cos^2 \theta}{\cos^2 \theta} \\
&= \frac{\sin^2 \theta + \cos^2 \theta}{\cos^2 \theta} \\
&= \frac{1}{\cos^2 \theta}
\end{aligned}
$$

(where we have used Property (1))

$$
= \sec^2 \theta.
$$

You Try It: Use identities (11) and (12) to calculate $\cos(\pi/12)$ and $\sin(\pi/12)$.

1.8 Sets and Functions

We have seen sets and functions throughout this review chapter, but it is well to bring out some of the ideas explicitly.

A *set* is a collection of objects. We denote a set with a capital roman letter, such as S or T or U. If S is a set and s is an object in that set then we write $s \in S$ and we say that *s is an element of* S. If S and T are sets then the collection of elements common to the two sets is called the *intersection* of S and T and is written $S \cap T$. The set of elements that are in S or in T or in both is called the *union* of S and T and is written $S \cup T$.

A *function* from a set S to a set T is a rule that assigns to each element of S a unique element of T. We write $f : S \to T$.

EXAMPLE 1.21

Let S be the set of all people who are alive at noon on October 10, 2004 and T the set of all real numbers. Let f be the rule that assigns to each person his or her weight in pounds at precisely noon on October 10, 2004. Discuss whether $f : S \to T$ is a function.

SOLUTION

Indeed f is a function since it assigns to each element of S a unique element of T. Notice that each person has just one weight at noon on October 10, 2004: that is a part of the definition of "function." However two different people *may have the same weight*—that is allowed.

EXAMPLE 1.22

Let S be the set of all people and T be the set of all people. Let f be the rule that assigns to each person his or her brother. Is f a function?

SOLUTION

In this case f is not a function. For many people have no brother (so the rule makes no sense for them) and many people have several brothers (so the rule is ambiguous for them).

EXAMPLE 1.23

Let S be the set of all people and T be the set of all strings of letters not exceeding 1500 characters (including blank spaces). Let f be the rule that assigns to each person his or her legal name. (Some people have rather long names; according to the *Guinness Book of World Records*, the longest has 1063 letters.) Determine whether $f : S \rightarrow T$ is a function.

SOLUTION

This f is a function because every person has one and only one legal name. Notice that several people may have the same name (such as "Jack Armstrong"), but that is allowed in the definition of function.

You Try It: Let f be the rule that assigns to each real number its cube root. Is this a function?

In calculus, the set S (called the *domain* of the function) and the set T (called the *range* of the function) will usually be sets of numbers; in fact they will often consist of one or more intervals in $\mathbb{R}$. The rule f will usually be given by one or several formulas. Many times the domain and range will not be given explicitly. These ideas will be illustrated in the examples below.

You Try It: Consider the rule that assigns to each real number its absolute value. Is this a function? Why or why not? If it is a function, then what are its domain and range?

1.8.1   EXAMPLES OF FUNCTIONS OF A REAL VARIABLE

EXAMPLE 1.24

Let $S = \mathbb{R}$, $T = \mathbb{R}$, and let $f(x) = x^2$. This is mathematical shorthand for the rule "assign to each $x \in S$ its square." Determine whether $f : \mathbb{R} \rightarrow \mathbb{R}$ is a function.

SOLUTION

We see that f is a function since it assigns to each element of S a unique element of T—namely its square.

Math Note: Notice that, in the definition of function, there is some imprecision in the definition of T. For instance, in Example 1.24, we could have let $T = [0, \infty)$ or $T = (-6, \infty)$ with no significant change in the function. In the example of the "name" function (Example 1.23), we could have let T be all strings of letters not exceeding 5000 characters in length. Or we could have made it all strings without regard to length. Likewise, in any of the examples we could make the set S smaller and the function would still make sense.

It is frequently convenient not to describe S and T explicitly.

EXAMPLE 1.25

Let $f(x) = +\sqrt{1 - x^2}$. Determine a domain and range for f which make f a function.

SOLUTION

Notice that f makes sense for $x \in [-1, 1]$ (we may not take the square root of a negative number, so we cannot allow $x > 1$ or $x < -1$). If we understand f to have domain $[-1, 1]$ and range $\mathbb{R}$, then $f : [-1, 1] \to \mathbb{R}$ is a function.

Math Note: When a function is given by a formula, as in Example 1.25, with no statement about the domain, then the domain is understood to be the set of all x for which the formula makes sense.

You Try It: Let

$$g(x) = \frac{x}{x^2 + 4x + 3}.$$

What are the domain and range of this function?

EXAMPLE 1.26

Let

$$f(x) = \begin{cases} -3 & \text{if } x \leq 1 \\ 2x^2 & \text{if } x > 1 \end{cases}$$

Determine whether f is a function.

SOLUTION

Notice that f unambiguously assigns to *each* real number another real number. The rule is given in two pieces, but it is still a valid rule. Therefore it is a function with domain equal to $\mathbb{R}$ and range equal to $\mathbb{R}$. It is also perfectly correct to take the range to be $(-4, \infty)$, for example, since f only takes values in this set.

Math Note: One point that you should learn from this example is that a function may be specified by *different formulas on different parts of the domain*.

You Try It: Does the expression

$$g(x) = \begin{cases} 4 & \text{if } x < 3 \\ x^2 - 7 & \text{if } x \geq 2 \end{cases}$$

define a function? Why or why not?

EXAMPLE 1.27

Let $f(x) = \pm\sqrt{x}$. Discuss whether f is a function.

SOLUTION

This f can only make sense for $x \geq 0$. But even then f is *not* a function since it is ambiguous. For instance, it assigns to $x = 1$ both the numbers 1 and -1.

1.8.2   GRAPHS OF FUNCTIONS

It is useful to be able to draw pictures which represent functions. These pictures, or *graphs*, are a device for helping us to think about functions. In this book we will only graph functions whose domains and ranges are subsets of the real numbers.

We graph functions in the x-y plane. The elements of the domain of a function are thought of as points of the x-axis. The values of a function are measured on the y-axis. The graph of f associates to x the unique y value that the function f assigns to x. In other words, a point (x, y) lies on the graph of f *if and only if* $y = f(x)$.

EXAMPLE 1.28

Let $f(x) = (x^2 + 2)/(x - 1)$. Determine whether there are points of the graph of f corresponding to $x = 3, 4$, and 1.

SOLUTION

The y value corresponding to $x = 3$ is $y = f(3) = 11/2$. Therefore the point $(3, 11/2)$ lies on the graph of f. Similarly, $f(4) = 6$ so that $(4, 6)$ lies on the graph. However, f is undefined at $x = 1$, so there is no point on the graph with x coordinate 1. The sketch in Fig. 1.38 was obtained by plotting several points.

Math Note: Notice that for each x in the domain of the function there is *one and only one* point on the graph—namely the unique point with y value equal to $f(x)$. If x is not in the domain of f, then there is no point on the graph that corresponds to x.

EXAMPLE 1.29

Is the curve in Fig. 1.39 the graph of a function?

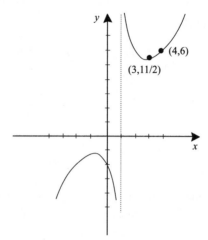

Fig. 1.38

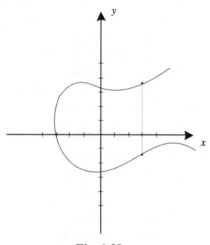

Fig. 1.39

SOLUTION

Observe that, corresponding to $x = 3$, for instance, there are two y values on the curve. Therefore the curve cannot be the graph of a function.

You Try It: Graph the function $y = x + |x|$.

EXAMPLE 1.30

Is the curve in Fig. 1.40 the graph of a function?

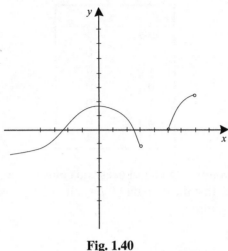

Fig. 1.40

SOLUTION

Notice that each x in the domain has just one y value corresponding to it. Thus, even though we cannot give a formula for the function, the curve is the graph of a function. The domain of this function is $(-\infty, 3) \cup (5, 7)$.

Math Note: A nice, geometrical way to think about the condition that each x in the domain has corresponding to it precisely one y value is this:

> If every vertical line drawn through a curve intersects that curve just once, then the curve is the graph of a function.

You Try It: Use the vertical line test to determine whether the locus $x^2 + y^2 = 1$ is the graph of a function.

1.8.3 PLOTTING THE GRAPH OF A FUNCTION

Until we learn some more sophisticated techniques, the basic method that we shall use for graphing functions is to plot points and then to connect them in a plausible manner.

EXAMPLE 1.31

Sketch the graph of $f(x) = x^3 - x$.

SOLUTION

We complete a table of values of the function f.

x	$y = x^3 - x$
-3	-24
-2	-6
-1	0
0	0
1	0
2	6
3	24

We plot these points on a pair of axes and connect them in a reasonable way (Fig. 1.41). Notice that the domain of f is all of $\mathbb{R}$, so we extend the graph to the edges of the picture.

EXAMPLE 1.32

Sketch the graph of

$$f(x) = \begin{cases} -1 & \text{if } x \leq 2 \\ x & \text{if } x > 2 \end{cases}$$

SOLUTION

We again start with a table of values.

x	$y = f(x)$
-3	-1
-2	-1
-1	-1
0	-1
1	-1
2	-1
3	3
4	4
5	5

We plot these on a pair of axes (Fig. 1.42).

Since the definition of the function changes at $x = 2$, we would be mistaken to connect these dots blindly. First notice that, for $x \leq 2$, the function is identically constant. Its graph is a horizontal line. For $x > 2$, the function is a line of slope 1. Now we can sketch the graph accurately (Fig. 1.43).

You Try It: Sketch the graph of $h(x) = |x| \cdot \sqrt[3]{x}$.

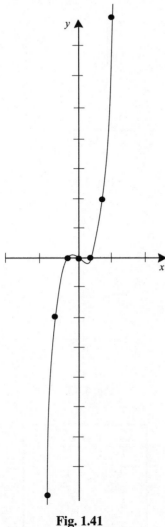

Fig. 1.41

EXAMPLE 1.33

Sketch the graph of $f(x) = \sqrt{x + 1}$.

SOLUTION

We begin by noticing that the domain of f, that is the values of x for which the function makes sense, is $\{x : x \geq -1\}$. The square root is understood to be the positive square root. Now we compute a table of values and plot some points.

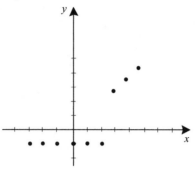

Fig. 1.42

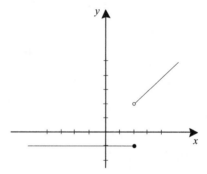

Fig. 1.43

x	$y = \sqrt{x+1}$
-1	0
0	1
1	$\sqrt{2}$
2	$\sqrt{3}$
3	2
4	$\sqrt{5}$
5	$\sqrt{6}$
6	$\sqrt{7}$

Connecting the points in a plausible way gives a sketch for the graph of f (Fig. 1.44).

EXAMPLE 1.34

Sketch the graph of $x = y^2$.

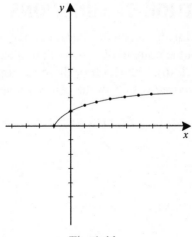

Fig. 1.44

SOLUTION

The sketch in Fig. 1.45 is obtained by plotting points. This curve is *not* the graph of a function.

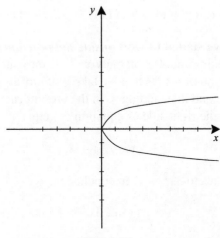

Fig. 1.45

A curve that is the plot of an equation but which is *not necessarily* the graph of a function is sometimes called the *locus* of the equation. When the curve *is* the graph of a function we usually emphasize this fact by writing the equation in the form $y = f(x)$.

You Try It: Sketch the locus $x = y^2 + y$.

1.8.4 COMPOSITION OF FUNCTIONS

Suppose that f and g are functions and that the domain of g contains the range of f. This means that if x is in the domain of f then $f(x)$ makes sense but also g may be applied to $f(x)$ (Fig. 1.46). The result of these two operations, one following the other, is called g *composed* with f or the *composition* of g with f. We write

$$(g \circ f)(x) = g(f(x)).$$

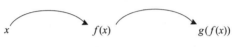

$$x \qquad f(x) \qquad g(f(x))$$

Fig. 1.46

EXAMPLE 1.35

Let $f(x) = x^2 - 1$ and $g(x) = 3x + 4$. Calculate $g \circ f$.

SOLUTION

We have

$$(g \circ f)(x) = g(f(x)) = g(x^2 - 1). \qquad (*)$$

Notice that we have started to work *inside the parentheses*: the first step was to substitute the definition of f, namely $x^2 - 1$, into our equation.

Now the definition of g says that we take g of *any* argument by multiplying that argument by 3 and then adding 4. In the present case we are applying g to $x^2 - 1$. Therefore the right side of equation $(*)$ equals

$$3 \cdot (x^2 - 1) + 4.$$

This easily simplifies to $3x^2 + 1$. In conclusion,

$$g \circ f(x) = 3x^2 + 1.$$

EXAMPLE 1.36

Let $f(t) = (t^2 - 2)/(t + 1)$ and $g(t) = 2t + 1$. Calculate $g \circ f$ and $f \circ g$.

SOLUTION

We calculate that

$$(g \circ f)(t) = g(f(t)) = g\left(\frac{t^2 - 2}{t + 1}\right). \qquad (**)$$

We compute g of any argument by doubling it and adding 1. Thus equation (∗∗) equals

$$2\left(\frac{t^2 - 2}{t + 1}\right) + 1$$
$$= \frac{2t^2 - 4}{t + 1} + 1$$
$$= \frac{2t^2 + t - 3}{t + 1}.$$

One of the main points of this example is to see that $f \circ g$ is different from $g \circ f$. We compute $f \circ g$:

$$(f \circ g)(t) = f(g(t))$$
$$= f(2t + 1)$$
$$= \frac{(2t + 1)^2 - 2}{(2t + 1) + 1}$$
$$= \frac{4t^2 + 4t - 1}{2t + 2}.$$

So $f \circ g$ and $g \circ f$ are different functions.

You Try It: Let $f(x) = |x|$ and $g(x) = \sqrt{x}/x$. Calculate $f \circ g(x)$ and $g \circ f(x)$.

We say a few words about *recognizing* compositions of functions.

EXAMPLE 1.37

How can we write the function $k(x) = (2x + 3)^2$ as the composition of two functions g and f?

SOLUTION

Notice that the function k can be thought of as two operations applied in sequence. First we double and add three, then we square. Thus define $f(x) = 2x + 3$ and $g(x) = x^2$. Then $k(x) = (g \circ f)(x)$.

We can also compose three (or more) functions: Define

$$(h \circ g \circ f)(x) = h(g(f(x))).$$

EXAMPLE 1.38

Write the function k from the last example as the composition of three functions (instead of just two).

SOLUTION

First we double, then we add 3, then we square. So let $f(x) = 2x$, $g(x) = x + 3$, $h(x) = x^2$. Then $k(x) = (h \circ g \circ f)(x)$.

EXAMPLE 1.39

Write the function

$$r(t) = \frac{2}{t^2 + 3}$$

as the composition of two functions.

SOLUTION

First we square t and add 3, then we divide 2 by the quantity just obtained. As a result, we define $f(t) = t^2 + 3$ and $g(t) = 2/t$. It follows that $r(t) = (g \circ f)(t)$.

You Try It: Express the function $g(x) = 3/(x^2 + 5)$ as the composition of two functions. Can you express it as the composition of three functions?

1.8.5   THE INVERSE OF A FUNCTION

Let f be the function which assigns to each working adult American his or her Social Security Number (a 9-digit string of integers). Let g be the function which assigns to each working adult American his or her age in years (an integer between 0 and 150). Both functions have the same domain, and both take values in the non-negative integers. But there is a fundamental difference between f and g. If you are given a Social Security number, then you can determine the person to whom it belongs. There will be one and only one person with that number. But if you are given a number between 0 and 150, then there will probably be millions of people with that age. You *cannot* identify a person by his/her age. In summary, if you know $g(x)$ then you generally *cannot* determine what x is. But if you know $f(x)$ then you *can* determine what (or who) x is. This leads to the main idea of this subsection.

Let $f : S \to T$ be a function. We say that f has an inverse (is invertible) if there is a function $f^{-1} : T \to S$ such that $(f \circ f^{-1})(t) = t$ for all $t \in T$ and $(f^{-1} \circ f)(s) = s$ for all $s \in S$. Notice that the symbol f^{-1} denotes a new function which we call the *inverse* of f.

Basic Rule for Finding Inverses   To find the inverse of a function f, we solve the equation

$$(f \circ f^{-1})(t) = t$$

for the function $f^{-1}(t)$.

EXAMPLE 1.40

Find the inverse of the function $f(s) = 3s$.

SOLUTION

We solve the equation

$$(f \circ f^{-1})(t) = t.$$

This is the same as

$$f(f^{-1}(t)) = t.$$

We can rewrite the last line as

$$3 \cdot f^{-1}(t) = t$$

or

$$f^{-1}(t) = \frac{t}{3}.$$

Thus $f^{-1}(t) = t/3$.

EXAMPLE 1.41

Let $f : \mathbb{R} \to \mathbb{R}$ be defined by $f(s) = 3s^5$. Find f^{-1}.

SOLUTION

We solve

$$(f \circ f^{-1})(t) = t$$

or

$$f(f^{-1}(t)) = t$$

or

$$3[f^{-1}(t)]^5 = t$$

or

$$[f^{-1}(t)]^5 = \frac{t}{3}$$

or

$$f^{-1}(t) = \left(\frac{t}{3}\right)^{1/5}.$$

You Try It: Find the inverse of the function $g(x) = \sqrt[3]{x} - 5$.

It is important to understand that some functions do *not* have inverses.

EXAMPLE 1.42

Let $f : \mathbb{R} \to \{t : t \geq 0\}$ be defined by $f(s) = s^2$. If possible, find f^{-1}.

SOLUTION

Using the Basic Rule, we attempt to solve

$$(f \circ f^{-1})(t) = t.$$

Writing this out, we have

$$[f^{-1}(t)]^2 = t.$$

But now there is a problem: we cannot solve this equation uniquely for $f^{-1}(t)$. We do not know whether $f^{-1}(t) = +\sqrt{t}$ or $f^{-1}(t) = -\sqrt{t}$. Thus f^{-1} is not a well defined function. Therefore f is not invertible and f^{-1} does not exist.

Math Note: There is a simple device which often enables us to obtain an inverse— even in situations like Example 1.42. We change the domain of the function. This idea is illustrated in the next example.

EXAMPLE 1.43

Define $\tilde{f} : \{s : s \geq 0\} \to \{t : t \geq 0\}$ by the formula $\tilde{f}(s) = s^2$. Find $\tilde{f}^{-1}$.

SOLUTION

We attempt to solve

$$(\tilde{f} \circ \tilde{f}^{-1})(t) = t.$$

Writing this out, we have

$$\tilde{f}(\tilde{f}^{-1}(t)) = t$$

or

$$[\tilde{f}^{-1}(t)]^2 = t.$$

This looks like the same situation we had in Example 1.42. But in fact things have improved. Now we *know* that $\tilde{f}^{-1}(t)$ *must be* $+\sqrt{t}$, because $\tilde{f}^{-1}$ must have range $S = \{s : s \geq 0\}$. Thus $\tilde{f}^{-1} : \{t : t \geq 0\} \to \{s : s \geq 0\}$ is given by $\tilde{f}^{-1}(t) = +\sqrt{t}$.

You Try It: The equation $y = x^2 + 3x$ does not describe the graph of an invertible function. Find a way to restrict the domain so that it is invertible.

Now we consider the graph of the inverse function. Suppose that $f : S \to T$ is invertible and that (s, t) is a point on the graph of f. Then $t = f(s)$ hence $s = f^{-1}(t)$ so that (t, s) is on the graph of f^{-1}. The geometrical connection between the points (s, t) and (t, s) is exhibited in Fig. 1.47: they are reflections of each other in the line $y = x$. We have discovered the following important principle:

The graph of f^{-1} is the reflection in the line $y = x$ of the graph of f.

Refer to Fig. 1.48.

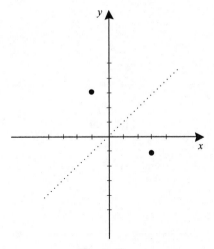

Fig. 1.47

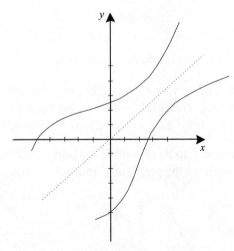

Fig. 1.48

EXAMPLE 1.44

Sketch the graph of the inverse of the function f whose graph is shown in Fig. 1.49.

SOLUTION

By inspection of the graph we see that f is one-to-one (i.e., takes different domain values to different range values) and onto (i.e., takes on all values in the range) from $S = [-2, 3]$ to $T = [1, 5]$. Therefore f has an inverse. The graph of f^{-1} is exhibited in Fig. 1.50.

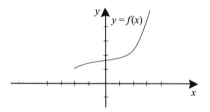

Fig. 1.49

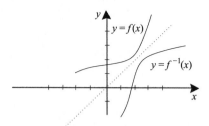

Fig. 1.50

You Try It: Sketch $f(x) = x^3 + x$ and its inverse.

Another useful fact is this: Since an invertible function must be one-to-one, two different x values cannot correspond to (that is, be "sent by the function to") the same y value. Looking at Figs. 1.51 and 1.52, we see that this means

In order for f to be invertible, no horizontal line can
intersect the graph of f more than once.

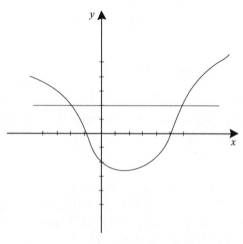

Fig. 1.51

In Fig. 1.51, the fact that the line $y = 2$ intersects the graph twice means that the function f takes the value 2 at two different points of its domain (namely at $x = -2$ and $x = 6$). Thus f is not one-to-one so it cannot be invertible. Figure 1.52 shows what happens if we try to invert f: the resulting curve is not the graph of a function.

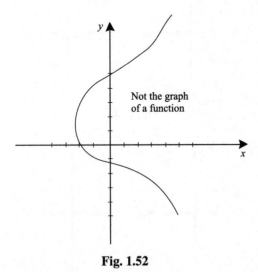

Not the graph
of a function

Fig. 1.52

EXAMPLE 1.45

Look at Figs. 1.53 and 1.55. Are the functions whose graphs are shown in parts (*a*) and (*b*) of each figure invertible?

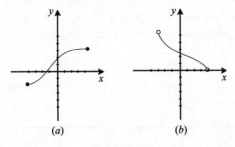

(*a*)   (*b*)

Fig. 1.53

SOLUTION

Graphs (*a*) and (*b*) in Fig. 1.53 are the graphs of invertible functions since no horizontal line intersects each graph more than once. Of course we must choose the domain and range appropriately. For (*a*) we take $S = [-4, 4]$ and $T = [-2, 3]$; for (*b*) we take $S = (-3, 4)$ and $T = (0, 5)$. Graphs (*a*) and (*b*) in Fig. 1.54 are the graphs of the inverse functions corresponding to (*a*) and (*b*) of Fig. 1.53 respectively. They are obtained by reflection in the line $y = x$.

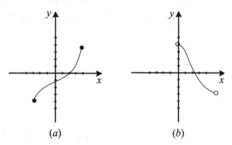

(*a*)   (*b*)

Fig. 1.54

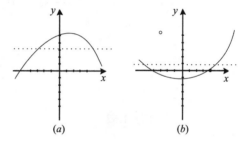

(*a*)   (*b*)

Fig. 1.55

In Fig. 1.55, graphs (*a*) and (*b*) are *not* the graphs of invertible functions. For each there is exhibited a horizontal line which intersects the graph twice. However graphs (*a*) and (*b*) in Fig. 1.56 exhibit a way to restrict the domains of the functions in (*a*) and (*b*) of Fig. 1.55 to make them invertible. Graphs (*a*) and (*b*) in Fig. 1.57 show their respective inverses.

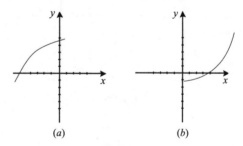

(*a*)   (*b*)

Fig. 1.56

You Try It: Give an example of a function from $\mathbb{R}$ to $\mathbb{R}$ that is not invertible, even when it is restricted to any interval of length 2.

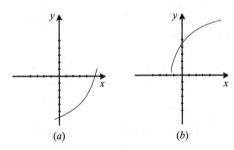

(a)   *(b)*

Fig. 1.57

1.9   A Few Words About Logarithms and Exponentials

We will give a more thorough treatment of the logarithm and exponential functions in Chapter 6. For the moment we record a few simple facts so that we may use these functions in the sections that immediately follow.

The logarithm is a function that is characterized by the property that

$$\log(x \cdot y) = \log x + \log y.$$

It follows from this property that

$$\log(x/y) = \log x - \log y$$

and

$$\log(x^n) = n \cdot \log x.$$

It is useful to think of $\log_a b$ as the power to which we raise a to get b, for any $a, b > 0$. For example, $\log_2 8 = 3$ and $\log_3(1/27) = -3$. This introduces the idea of the *logarithm to a base*.

You Try It: Calculate $\log_5 125$, $\log_3(1/81)$, $\log_2 16$.

The most important base for the logarithm is Euler's number $e \approx 2.71828\ldots$. Then we write $\ln x = \log_e x$. For the moment we take the logarithm to the base e, or the *natural logarithm*, to be given. It is characterized among all logarithm functions by the fact that its graph has tangent line with slope 1 at $x = 1$. See Fig. 1.58. Then we set

$$\log_a x = \frac{\ln x}{\ln a}.$$

Note that this formula gives immediately that $\log_e x = \ln x$, once we accept that $\log_e e = 1$.

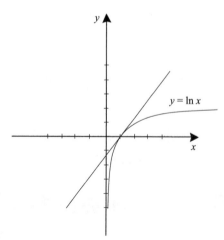

Fig. 1.58

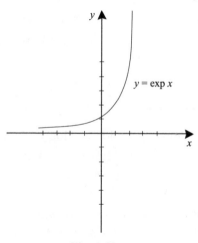

Fig. 1.59

Math Note: In mathematics, we commonly write $\log x$ to mean the natural logarithm. Thus you will sometimes encounter $\ln x$ and sometimes encounter $\log x$ (without any subscript); they are both understood to mean $\log_e x$, the natural logarithm.

The exponential function $\exp x$ is defined to be the inverse function to $\ln x$. Figure 1.59 shows the graph of $y = \exp x$. In fact we will see later that $\exp x = e^x$. More generally, the function a^x is the inverse function to $\log_a x$. The exponential has these properties:

(a) $a^{b+c} = a^b \cdot a^c$;

(b) $(a^b)^c = a^{b \cdot c}$;

(c) $a^{b-c} = \dfrac{a^b}{a^c}$.

These are really just restatements of properties of the logarithm function that we have already considered.

You Try It: Simplify the expressions $3^2 \cdot 5^4/(15)^3$ and $2^4 \cdot 6^3 \cdot 12^{-4}$.

Exercises

1. Each of the following is a rational number. Write it as the quotient of two integers.

  (a) $2/3 - 7/8$

  (b) 43.219445

  (c) $\dfrac{-37}{533} \cdot \dfrac{-4}{-6}$

  (d) $\dfrac{2}{3.45969696\ldots}$

  (e) $-73.235677677677\ldots$

  (f) $\dfrac{\frac{3}{5}}{\frac{-17}{4} + \frac{3}{9}}$

  (g) $\dfrac{\frac{-4}{9} + \frac{2}{5}}{\frac{-11}{3} + \frac{6}{7}}$

  (h) $3.2147569569569\ldots$

2. Plot the numbers $3.4, -\pi/2, 2\pi, -\sqrt{2}+1, \sqrt{3} \cdot 4, 9/2, -29/10$ on a real number line. Label each plotted point.

3. Sketch each of the following sets on a separate real number line.

  (a) $S = \{x \in \mathbb{R}: |x - 2| < 4\}$

  (b) $T = \{t \in \mathbb{R}: t^2 + 1 = 5\}$

  (c) $U = \{s \in \mathbb{R}: 2s - 5 \le 3\}$

  (d) $V = \{y \in \mathbb{R}: |6y + 1| > 2\}$

  (e) $S = \{x \in \mathbb{R}: x^2 + 3 < 6\}$

  (f) $T = \{s \in \mathbb{R}: |s| = |s + 1|\}$

4. Plot each of the points $(2, -4)$, $(-6, 3)$, (π, π^2), $(-\sqrt{5}, \sqrt{8})$, $(\sqrt{2\pi}, -3)$, $(1/3, -19/4)$ on a pair of cartesian coordinate axes. Label each point.

5. Plot each of these planar loci on a separate set of axes.

   (a) $\{(x, y): y = 2x^2 - 3\}$
   (b) $\{(x, y): x^2 + y^2 = 9\}$
   (c) $y = x^3 + x$
   (d) $x = y^3 + y$
   (e) $x = y^2 - y^3$
   (f) $x^2 + y^4 = 3$

6. Plot each of these regions in the plane.

   (a) $\{(x, y): x^2 + y^2 < 4\}$
   (b) $\{(x, y): y > x^2\}$
   (c) $\{(x, y): y < x^3\}$
   (d) $\{(x, y): x \geq 2y + 3\}$
   (e) $\{(x, y): y \leq x + 1\}$
   (f) $\{(x, y): 2x + y \geq 1\}$

7. Calculate the slope of each of the following lines:

   (a) The line through the points $(-5, 6)$ and $(2, 4)$
   (b) The line perpendicular to the line through $(1, 2)$ and $(3, 4)$
   (c) The line $2y + 3x = 6$
   (d) The line $\dfrac{x - 4y}{x + y} = 6$
   (e) The line through the points $(1, 1)$ and $(-8, 9)$
   (f) The line $x - y = 4$

8. Write the equation of each of the following lines.

   (a) The line parallel to $3x + 8y = -9$ and passing through the point $(4, -9)$.
   (b) The line perpendicular to $x + y = 2$ and passing through the point $(-4, -8)$.
   (c) The line passing through the point $(4, 6)$ and having slope -8.
   (d) The line passing through $(-6, 4)$ and $(2, 3)$.
   (e) The line passing through the origin and having slope 6.
   (f) The line perpendicular to $x = 3y - 7$ and passing through $(-4, 7)$.

9. Graph each of the lines in Exercise 8 on its own set of axes. Label your graphs.

10. Which of the following is a function and which is not? Give a reason in each case.

(a) f assigns to each person his biological father
(b) g assigns to each man his dog
(c) h assigns to each real number its square root
(d) f assigns to each positive integer its cube
(e) g assigns to each car its driver
(f) h assigns to each toe its foot
(g) f assigns to each rational number the greatest integer that does not exceed it
(h) g assigns to each integer the next integer
(i) h assigns to each real number its square plus six

11.   Graph each of these functions on a separate set of axes. Label your graph.

(a) $f(x) = 3x^2 - x$
(b) $g(x) = \dfrac{x+2}{x}$
(c) $h(x) = x^3 - x^2$
(d) $f(x) = 3x + 2$
(e) $g(x) = x^2 - 2x$
(f) $h(x) = \sqrt{x} + 3$

12.   Calculate each of the following trigonometric quantities.

(a) $\sin(8\pi/3)$
(b) $\tan(-5\pi/6)$
(c) $\sec(7\pi/4)$
(d) $\csc(13\pi/4)$
(e) $\cot(-15\pi/4)$
(f) $\cos(-3\pi/4)$

13.   Calculate the left and right sides of the twelve fundamental trigonometric identities for the values $\theta = \pi/3$ and $\psi = -\pi/6$, thus confirming the identities for these particular values.

14.   Sketch the graphs of each of the following trigonometric functions.

(a) $f(x) = \sin 2x$
(b) $g(x) = \cos(x + \pi/2)$
(c) $h(x) = \tan(-x + \pi)$
(d) $f(x) = \cot(3x + \pi)$
(e) $g(x) = \sin(x/3)$
(f) $h(x) = \cos(-\pi + [x/2])$

15.   Convert each of the following angles *from radian measure* to *degree measure*.

(a) $\theta = \pi/24$
(b) $\theta = -\pi/3$
(c) $\theta = 27\pi/12$
(d) $\theta = 9\pi/16$
(e) $\theta = 3$
(f) $\theta = -5$

16. Convert each of the following angles *from degree measure* to *radian measure*.

(a) $\theta = 65°$
(b) $\theta = 10°$
(c) $\theta = -75°$
(d) $\theta = -120°$
(e) $\theta = \pi°$
(f) $\theta = 3.14°$

17. For each of the following pairs of functions, calculate $f \circ g$ and $g \circ f$.

(a) $f(x) = x^2 + 2x + 3$ $g(x) = (x-1)^2$
(b) $f(x) = \sqrt{x+1}$ $g(x) = \sqrt[3]{x^2 - 2}$
(c) $f(x) = \sin(x + 3x^2)$ $g(x) = \cos(x^2 - x)$
(d) $f(x) = e^{x+2}$ $g(x) = \ln(x - 5)$
(e) $f(x) = \sin(x^2 + x)$ $g(x) = \ln(x^2 - x)$
(f) $f(x) = e^{x^2}$ $g(x) = e^{-x^2}$
(g) $f(x) = x(x+1)(x+2)$ $g(x) = (2x-3)(x+4)$

18. Consider each of the following as functions from $\mathbb{R}$ to $\mathbb{R}$ and say whether the function is invertible. If it is, find the inverse with an explicit formula.

(a) $f(x) = x^3 + 5$
(b) $g(x) = x^2 - x$
(c) $h(x) = (\operatorname{sgn} x) \cdot \sqrt{|x|}$, where $\operatorname{sgn} x$ is $+1$ if x is positive, -1 if x is negative, 0 if x is 0.
(d) $f(x) = x^5 + 8$
(e) $g(x) = e^{-3x}$
(f) $h(x) = \sin x$
(g) $f(x) = \tan x$
(h) $g(x) = (\operatorname{sgn} x) \cdot x^2$, where $\operatorname{sgn} x$ is $+1$ if x is positive, -1 if x is negative, 0 if x is 0.

19. For each of the functions in Exercise 18, graph both the function and its inverse in the same set of axes.

20. Determine whether each of the following functions, on the given domain S, is invertible. If it is, then find the inverse explicitly.

(a) $f(x) = x^2,$ $S = [2, 7]$

(b) $g(x) = \ln x,$ $S = [1, \infty)$

(c) $h(x) = \sin x,$ $S = [0, \pi/2]$

(d) $f(x) = \cos x,$ $S = [0, \pi]$

(e) $g(x) = \tan x,$ $S = (-\pi/2, \pi/2)$

(f) $h(x) = x^2,$ $S = [-2, 5]$

(g) $f(x) = x^2 - 3x,$ $S = [4, 7]$

CHAPTER 2

Foundations of Calculus

2.1   Limits

The single most important idea in calculus is the idea of limit. More than 2000 years ago, the ancient Greeks wrestled with the limit concept, and *they did not succeed*. It is only in the past 200 years that we have finally come up with a firm understanding of limits. Here we give a brief sketch of the essential parts of the limit notion.

Suppose that f is a function whose domain contains two neighboring intervals: $f : (a, c) \cup (c, b) \to \mathbb{R}$. We wish to consider the behavior of f as the variable x approaches c. If $f(x)$ approaches a particular finite value ℓ as x approaches c, then we say that *the function f has the limit ℓ as x approaches c*. We write

$$\lim_{x \to c} f(x) = \ell.$$

The rigorous mathematical definition of limit is this:

Definition 2.1   Let $a < c < b$ and let f be a function whose domain contains $(a, c) \cup (c, b)$. We say that f has limit ℓ at c, and we write $\lim_{x \to c} f(x) = \ell$ when this condition holds: For each $\epsilon > 0$ there is a $\delta > 0$ such that

$$|f(x) - \ell| < \epsilon$$

whenever $0 < |x - c| < \delta$.

It is important to know that there is a rigorous definition of the limit concept, and any development of mathematical theory relies in an essential way on this rigorous definition. However, in the present book we may make good use of an intuitive understanding of limit. We now develop that understanding with some carefully chosen examples.

EXAMPLE 2.1

Define

$$f(x) = \begin{cases} 3 - x & \text{if } x < 1 \\ x^2 + 1 & \text{if } x > 1 \end{cases}$$

See Fig. 2.1. Calculate $\lim_{x \to 1} f(x)$.

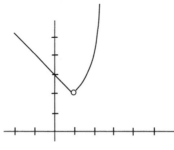

Fig. 2.1

SOLUTION

Observe that, when x is to the left of 1 and very near to 1 then $f(x) = 3 - x$ is very near to 2. Likewise, when x is to the right of 1 and very near to 1 then $f(x) = x^2 + 1$ is very near to 2. We conclude that

$$\lim_{x \to 1} f(x) = 2.$$

We have successfully calculated our first limit. Figure 2.1 confirms the conclusion that our calculations derived.

EXAMPLE 2.2

Define

$$g(x) = \frac{x^2 - 4}{x - 2}.$$

Calculate $\lim_{x \to 2} g(x)$.

SOLUTION

We observe that both the numerator and the denominator of the fraction defining g tend to 0 as $x \to 2$ (i.e., as x tends to 2). Thus the question seems to be indeterminate.

However, we may factor the numerator as $x^2 - 4 = (x - 2)(x + 2)$. As long as $x \neq 2$ (and these are the only x that we examine when we calculate $\lim_{x \to 2}$), we can then divide the denominator of the expression defining g into the numerator. Thus

$$g(x) = x + 2 \qquad \text{for } x \neq 2.$$

Now

$$\lim_{x \to 2} g(x) = \lim_{x \to 2} x + 2 = 4.$$

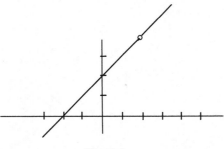

Fig. 2.2

The graph of the function g is shown in Fig. 2.2. We encourage the reader to use a pocket calculator to calculate values of g for x near 2 but unequal to 2 to check the validity of our answer. For example,

x	$g(x) = [x^2 - 4]/[x - 2]$
1.8	3.8
1.9	3.9
1.99	3.99
1.999	3.999
2.001	4.001
2.01	4.01
2.1	4.1
2.2	4.2

We see that, when x is close to 2 (but unequal to 2), then $g(x)$ is close (indeed, as close as we please) to 4.

You Try It: Calculate the limit $\lim_{x \to 3} \dfrac{x^3 - 3x^2 + x - 3}{x - 3}$.

Math Note: It must be stressed that, when we calculate $\lim_{x \to c} f(x)$, we *do not* evaluate f at c. In the last example it would have been impossible to do so. We want to determine what we *anticipate* f will do as x approaches c, *not* what value (if any) f *actually takes* at c. The next example illustrates this point rather dramatically.

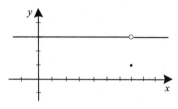

Fig. 2.3

EXAMPLE 2.3

Define

$$h(x) = \begin{cases} 3 & \text{if } x \neq 7 \\ 1 & \text{if } x = 7 \end{cases}$$

Calculate $\lim_{x \to 7} h(x)$.

SOLUTION

It would be incorrect to simply plug the value 7 into the function h and thereby to conclude that the limit is 1. In fact when x is *near to* 7 but *unequal to* 7, we see that h takes the value 3. This statement is true *no matter how close x is to* 7. We conclude that $\lim_{x \to 7} h(x) = 3$.

You Try It: Calculate $\lim_{x \to 4} [x^2 - x - 12]/[x - 4]$.

2.1.1 ONE-SIDED LIMITS

There is also a concept of one-sided limit. We say that

$$\lim_{x \to c^-} f(x) = \ell$$

if the values of f become closer and closer to ℓ when x is near to c *but on the left*. In other words, in studying $\lim_{x \to c^-} f(x)$, we only consider values of x that are less than c.

Likewise, we say that

$$\lim_{x \to c^+} f(x) = \ell$$

if the values of f become closer and closer to ℓ when x is near to c *but on the right*. In other words, in studying $\lim_{x \to c^+} f(x)$, we only consider values of x that are greater than c.

EXAMPLE 2.4

Discuss the limits of the function

$$f(x) = \begin{cases} 2x - 4 & \text{if } x < 2 \\ x^2 & \text{if } x \geq 2 \end{cases}$$

at $c = 2$.

SOLUTION

As x approaches 2 from the left, $f(x) = 2x - 4$ approaches 0. As x approaches 2 from the right, $f(x) = x^2$ approaches 4. Thus we see that f has left limit 0 at $c = 2$, written

$$\lim_{x \to 2^-} f(x) = 0,$$

and f has right limit 4 at $c = 2$, written

$$\lim_{x \to 2^+} f(x) = 4.$$

Note that the full limit $\lim_{x \to 2} f(x)$ *does not exist* (because the left and right limits are unequal).

You Try It: Discuss one-sided limits at $c = 3$ for the function

$$f(x) = \begin{cases} x^3 - x & \text{if } x < 3 \\ 24 & \text{if } x = 3 \\ 4x + 1 & \text{if } x > 3 \end{cases}$$

All the properties of limits that will be developed in this chapter, as well as the rest of the book, apply equally well to one-sided limits as to two-sided (or standard) limits.

2.2   Properties of Limits

To increase our facility in manipulating limits, we have certain arithmetical and functional rules about limits. Any of these may be verified using the rigorous definition of limit that was provided at the beginning of the last section. We shall state the rules and get right to the examples.

If f and g are two functions, c is a real number, and $\lim_{x \to c} f(x)$ and $\lim_{x \to c} g(x)$ exist, then

Theorem 2.1

(a) $\lim_{x \to c}(f \pm g)(x) = \lim_{x \to c} f(x) \pm \lim_{x \to c} g(x);$

(b) $\lim_{x \to c} (f \cdot g)(x) = (\lim_{x \to c} f(x)) \cdot (\lim_{x \to c} g(x))$;

(c) $\lim_{x \to c} \left(\dfrac{f}{g} \right)(x) = \dfrac{\lim_{x \to c} f(x)}{\lim_{x \to c} g(x)}$ *provided that* $\lim_{x \to c} g(x) \neq 0$;

(d) $\lim_{x \to c} (\alpha \cdot f(x)) = \alpha \cdot (\lim_{x \to c} f(x))$ *for any constant* α.

Some theoretical results, which will prove useful throughout our study of calculus, are these:

Theorem 2.2
Let $a < c < b$. A function f on the interval $\{x : a < x < b\}$ cannot have two distinct limits at c.

Theorem 2.3
If

$$\lim_{x \to c} g(x) = 0$$

and

$$\lim_{x \to c} f(x) \text{ either does not exist or exists and is not zero}$$

then

$$\lim_{x \to c} \frac{f(x)}{g(x)}$$

does not exist.

Theorem 2.4 (The Pinching Theorem)
Suppose that $f, g,$ and h are functions whose domains each contain $S = (a, c) \cup (c, b)$. Assume further that

$$g(x) \leq f(x) \leq h(x)$$

for all $x \in S$. Refer to Fig. 2.4.

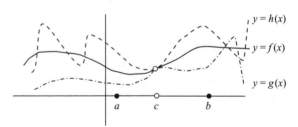

Fig. 2.4

If

$$\lim_{x \to c} g(x) = \ell$$

and

$$\lim_{x \to c} h(x) = \ell$$

then

$$\lim_{x \to c} f(x) = \ell.$$

EXAMPLE 2.5

Calculate $\lim_{x \to 3} 4x^3 - 7x^2 + 5x - 9$.

SOLUTION

We may apply Theorem 2.1(a) repeatedly to see that

$$\lim_{x \to 3} 4x^3 - 7x^2 + 5x - 9 = \lim_{x \to 3} 4x^3 - \lim_{x \to 3} 7x^2 + \lim_{x \to 3} 5x - \lim_{x \to 3} 9. \qquad (*)$$

We next observe that $\lim_{x \to 3} x = 3$. This assertion is self-evident, for when x is near to 3 then x is near to 3. Applying Theorem 2.1(d) and Theorem 2.1(b) repeatedly, we now see that

$$\lim_{x \to 3} 4x^3 = 4 \cdot \left[\lim_{x \to 3} x \right] \cdot \left[\lim_{x \to 3} x \right] \cdot \left[\lim_{x \to 3} x \right] = 4 \cdot 3 \cdot 3 \cdot 3 = 108.$$

Also

$$\lim_{x \to 3} 7x^2 = 7 \cdot \left[\lim_{x \to 3} x \right] \cdot \left[\lim_{x \to 3} x \right] = 7 \cdot 3 \cdot 3 = 63,$$

$$\lim_{x \to 3} 5x = 5 \cdot \left[\lim_{x \to 3} x \right] = 5 \cdot 3 = 15.$$

Of course $\lim_{x \to 3} 9 = 9$.

Putting all this information into equation $(*)$ gives

$$\lim_{x \to 3} 4x^3 - 7x^2 + 5x - 9 = 108 - 63 + 15 - 9 = 51.$$

EXAMPLE 2.6

Use the Pinching Theorem to analyze the limit

$$\lim_{x \to 0} x \sin x.$$

SOLUTION

We observe that

$$-|x| \equiv g(x) \le f(x) = x \sin x \le h(x) \equiv |x|.$$

Thus we may apply the Pinching Theorem. Obviously

$$\lim_{x \to 0} g(x) = \lim_{x \to 0} h(x) = 0.$$

We conclude that $\lim_{x \to 0} f(x) = 0$.

EXAMPLE 2.7

Analyze the limit

$$\lim_{x \to -2} \frac{x^2 + 4}{x + 2}.$$

SOLUTION

The denominator tends to 0 while the numerator does not. According to Theorem 2.3, the limit cannot exist.

You Try It: Use the Pinching Theorem to calculate $\lim_{x \to 0} x^2 \sin x$.

You Try It: What can you say about $\lim_{x \to -1} \dfrac{x^2}{x^2 - 1}$?

2.3   Continuity

Let f be a function whose domain contains the interval (a, b). Assume that c is a point of (a, b). We say that the function f is *continuous* at c if

$$\lim_{x \to c} f(x) = f(c).$$

Conceptually, f is continuous at c if *the expected value of f at c equals the actual value of f at c.*

EXAMPLE 2.8

Is the function

$$f(x) = \begin{cases} 2x^2 - x & \text{if } x < 2 \\ 3x & \text{if } x \geq 2 \end{cases}$$

continuous at $x = 2$?

SOLUTION

We easily check that $\lim_{x \to 2} f(x) = 6$. Also the actual value of f at 2, given by the second part of the formula, is equal to 6. By the definition of continuity, we may conclude that f is continuous at $x = 2$. See Fig. 2.5.

EXAMPLE 2.9

Where is the function

$$g(x) = \begin{cases} \dfrac{1}{x - 3} & \text{if } x < 4 \\ 2x + 3 & \text{if } x \geq 4 \end{cases}$$

continuous?

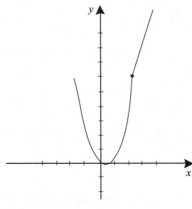

Fig. 2.5

SOLUTION

If $x < 3$ then the function is plainly continuous. The function is undefined at $x = 3$ so we may not even speak of continuity at $x = 3$. The function is also obviously continuous for $3 < x < 4$. At $x = 4$ the limit of g does not exist—it is 1 from the left and 11 from the right. So the function is not continuous (we sometimes say that it is *discontinuous*) at $x = 4$. By inspection, the function is continuous for $x > 4$.

You Try It: Discuss continuity of the function

$$g(x) = \begin{cases} x - x^2 & \text{if } x < -2 \\ 10 & \text{if } x = -2 \\ -5x & \text{if } x > -2 \end{cases}$$

We note that Theorem 2.1 guarantees that the collection of continuous functions is closed under addition, subtraction, multiplication, division (as long as we do not divide by 0), and scalar multiplication.

Math Note: If $f \circ g$ makes sense, if $\lim_{x \to c} g(x) = \ell$, and if $\lim_{s \to \ell} f(s) = m$, then it does not necessarily follow that $\lim_{x \to c} f \circ g(x) = m$. [We invite the reader to find an example.] One must assume, in addition, that f is continuous at ℓ. This point will come up from time to time in our later studies.

In the next section we will learn the concept of the derivative. It will turn out that a function that possesses the derivative is also continuous.

2.4 The Derivative

Suppose that f is a function whose domain contains the interval (a, b). Let c be a point of (a, b). If the limit

$$\lim_{h \to 0} \frac{f(c + h) - f(c)}{h} \qquad (*)$$

exists then we say that f is differentiable at c and we call the limit the derivative of f at c.

EXAMPLE 2.10

Is the function $f(x) = x^2 + x$ differentiable at $x = 2$? If it is, calculate the derivative.

SOLUTION

We calculate the limit $(*)$, with the role of c played by 2:

$$\lim_{h \to 0} \frac{f(2 + h) - f(2)}{h} = \lim_{h \to 0} \frac{[(2 + h)^2 + (2 + h)] - [2^2 + 2]}{h}$$

$$= \lim_{h \to 0} \frac{[(4 + 4h + h^2) + (2 + h)] - [6]}{h}$$

$$= \lim_{h \to 0} \frac{5h + h^2}{h}$$

$$= \lim_{h \to 0} 5 + h$$

$$= 5.$$

We see that the required limit $(*)$ exists, and that it equals 5. Thus the function $f(x) = x^2 + x$ is differentiable at $x = 2$, and the value of the derivative is 5.

Math Note: When the derivative of a function f exists at a point c, then we denote the derivative either by $f'(c)$ or by $(d/dx)f(c) = (df/dx)(c)$. In some contexts (e.g., physics) the notation $\dot{f}(c)$ is used. In the last example, we calculated that $f'(2) = 5$.

The importance of the derivative is two-fold: it can be interpreted as *rate of change* and it can be interpreted as the *slope*. Let us now consider both of these ideas.

Suppose that $\varphi(t)$ represents the position (in inches or feet or some other standard unit) of a moving body at time t. At time 0 the body is at $\varphi(0)$, at time 3 the body is at $\varphi(3)$, and so forth. Imagine that we want to determine the *instantaneous velocity* of the body at time $t = c$. What could this mean? One reasonable interpretation is that we can calculate the average velocity over a small interval at c, and let the length of that interval shrink to zero to determine the instantaneous velocity. To carry out this program, imagine a short interval $[c, c + h]$. The *average velocity* of the moving body over that interval is

$$v_{av} \equiv \frac{\varphi(c + h) - \varphi(c)}{h}.$$

This is a familiar expression (see (∗)). As we let $h \to 0$, we know that this expression tends to the derivative of φ at c. On the other hand, it is reasonable to declare this limit to be the instantaneous velocity. We have discovered the following important rule:

> Let φ be a differentiable function on an interval (a, b). Suppose that $\varphi(t)$ represents the position of a moving body. Let $c \in (a, b)$. Then
>
> $\varphi'(c) =$ instantaneous velocity of the moving body at c.

Now let us consider slope. Look at the graph of the function $y = f(x)$ in Fig. 2.6. We wish to determine the "slope" of the graph at the point $x = c$. This is the same as determining the slope of the *tangent line* to the graph of f at $x = c$, where the tangent line is the line that best approximates the graph at that point. See Fig. 2.7. What could this mean? After all, it takes two points to determine the slope of a line, yet we are only given the point $(c, f(c))$ on the graph. One reasonable interpretation of the slope at $(c, f(c))$ is that it is the limit of the slopes of secant lines determined

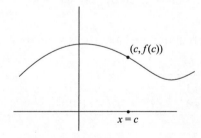

Fig. 2.6

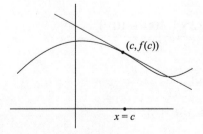

Fig. 2.7

by $(c, f(c))$ and nearby points $(c+h, f(c+h))$. See Fig. 2.8. Let us calculate this limit:

$$\lim_{h \to 0} \frac{f(c+h) - f(c)}{(c+h) - c} = \lim_{h \to 0} \frac{f(c+h) - f(c)}{h}.$$

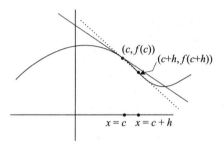

Fig. 2.8

We know that this last limit (the same as (∗)) is the derivative of f at c. We have learned the following:

Let f be a differentiable function on an interval (a, b). Let $c \in (a, b)$. Then the slope of the tangent line to the graph of f at c is $f'(c)$.

EXAMPLE 2.11

Calculate the instantaneous velocity at time $t = 5$ of an automobile whose position at time t seconds is given by $g(t) = t^3 + 4t^2 + 10$ feet.

SOLUTION

We know that the required instantaneous velocity is $g'(5)$. We calculate

$$
\begin{aligned}
g'(5) &= \lim_{h \to 0} \frac{g(5+h) - g(5)}{h} \\
&= \lim_{h \to 0} \frac{[(5+h)^3 + 4(5+h)^2 + 10] - [5^3 + 4 \cdot 5^2 + 10]}{h} \\
&= \lim_{h \to 0} \left[\frac{((125 + 75h + 15h^2 + h^3) + 4 \cdot (25 + 10h + h^2) + 10)}{h} \right. \\
&\qquad\qquad \left. - \frac{(125 + 100 + 10)}{h} \right] \\
&= \lim_{h \to 0} \frac{115h + 19h^2 + h^3}{h} \\
&= \lim_{h \to 0} 115 + 19h + h^2 \\
&= 115.
\end{aligned}
$$

We conclude that the instantaneous velocity of the moving body at time $t = 5$ is $g'(5) = 115$ ft/sec.

Math Note: Since position (or distance) is measured in feet, and time in seconds, then we measure velocity in feet per second.

EXAMPLE 2.12

Calculate the slope of the tangent line to the graph of $y = f(x) = x^3 - 3x$ at $x = -2$. Write the equation of the tangent line. Draw a figure illustrating these ideas.

SOLUTION

We know that the desired slope is equal to $f'(-2)$. We calculate

$$
\begin{aligned}
f'(-2) &= \lim_{h \to 0} \frac{f(-2 + h) - f(-2)}{h} \\
&= \lim_{h \to 0} \frac{[(-2 + h)^3 - 3(-2 + h)] - [(-2)^3 - 3(-2)]}{h} \\
&= \lim_{h \to 0} \frac{[(-8 + 12h - 6h^2 + h^3) + (6 - 3h)] + [2]}{h} \\
&= \lim_{h \to 0} \frac{h^3 - 6h^2 + 9h}{h} \\
&= \lim_{h \to 0} h^2 - 6h + 9 \\
&= 9.
\end{aligned}
$$

We conclude that the slope of the tangent line to the graph of $y = x^3 - 3x$ at $x = -2$ is 9. The tangent line passes through $(-2, f(-2)) = (-2, -2)$ and has slope 9. Thus it has equation

$$y - (-2) = 9(x - (-2)).$$

The graph of the function and the tangent line are exhibited in Fig. 2.9.

You Try It: Calculate the tangent line to the graph of $f(x) = 4x^2 - 5x + 2$ at the point where $x = 2$.

EXAMPLE 2.13

A rubber balloon is losing air steadily. At time t minutes the balloon contains $75 - 10t^2 + t$ cubic inches of air. What is the rate of loss of air in the balloon at time $t = 1$?

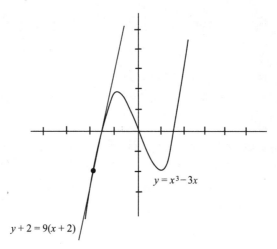

$y = x^3 - 3x$

$y + 2 = 9(x + 2)$

Fig. 2.9

SOLUTION

Let $\psi(t) = 75 - 10t^2 + t$. Of course the rate of loss of air is given by $\psi'(1)$. We therefore calculate

$$\psi'(1) = \lim_{h \to 0} \frac{\psi(1+h) - \psi(1)}{h}$$

$$= \lim_{h \to 0} \frac{[75 - 10(1+h)^2 + (1+h)] - [75 - 10 \cdot 1^2 + 1]}{h}$$

$$= \lim_{h \to 0} \frac{[75 - (10 + 20h + 10h^2) + (1+h)] - [66]}{h}$$

$$= \lim_{h \to 0} \frac{-19h - 10h^2}{h}$$

$$= \lim_{h \to 0} -19 - 10h$$

$$= -19.$$

In conclusion, the rate of air loss in the balloon at time $t = 1$ is $\psi'(1) = -19\,\text{ft}^3/\text{sec}$. Observe that the negative sign in this answer indicates that the change is *negative*, i.e., that the quantity is decreasing.

You Try It: The amount of water in a leaky tank is given by $W(t) = 50 - 5t^2 + t$ gallons. What is the rate of leakage of the water at time $t = 2$?

Math Note: We have noted that the derivative may be used to describe a rate of change and also to denote the slope of the tangent line to a graph. These are really two different manifestations of the same thing, for a slope is the rate of change of rise with respect to run (see Section 1.4 on the slope of a line).

2.5 Rules for Calculating Derivatives

Calculus is a powerful tool, for much of the physical world that we wish to analyze is best understood in terms of rates of change. It becomes even more powerful when we can find some simple rules that enable us to calculate derivatives quickly and easily. This section is devoted to that topic.

I Derivative of a Sum [The Sum Rule]:   We calculate the derivative of a sum (or difference) by

$$(f(x) \pm g(x))' = f'(x) \pm g'(x).$$

In our many examples, we have used this fact implicitly. We are now just enunciating it formally.

II Derivative of a Product [The Product Rule]:   We calculate the derivative of a product by

$$[f(x) \cdot g(x)]' = f'(x) \cdot g(x) + f(x) \cdot g'(x).$$

We urge the reader to test this formula on functions that we have worked with before. It has a surprising form. Note in particular that it is *not* the case that $[f(x) \cdot g(x)]' = f'(x) \cdot g'(x)$.

III Derivative of a Quotient [The Quotient Rule]:   We calculate the derivative of a quotient by

$$\left[\frac{f(x)}{g(x)} \right]' = \frac{g(x) \cdot f'(x) - f(x) \cdot g'(x)}{g^2(x)}.$$

In fact one can derive this new formula by applying the product formula to $g(x) \cdot [f(x)/g(x)]$. We leave the details for the interested reader.

IV Derivative of a Composition [The Chain Rule]:   We calculate the derivative of a composition by

$$[f \circ g(x)]' = f'(g(x)) \cdot g'(x).$$

To make optimum use of these four new formulas, we need a library of functions to which to apply them.

A Derivatives of Powers of x: If $f(x) = x^k$ then $f'(x) = k \cdot x^{k-1}$, where $k \in \{0, 1, 2, \ldots\}$.

Math Note: If you glance back at the examples we have done, you will notice that we have already calculated that the derivative of x is 1, the derivative of x^2 is $2x$, and the derivative of x^3 is $3x^2$. The rule just enunciated is a generalization of these facts, and is established in just the same way.

B   Derivatives of Trigonometric Functions: The rules for differentiating sine and cosine are simple and elegant:

1. $\dfrac{d}{dx} \sin x = \cos x.$

2. $\dfrac{d}{dx} \cos x = -\sin x.$

We can find the derivatives of the other trigonometric functions by using these two facts together with the quotient rule from above:

3. $\dfrac{d}{dx} \tan x = \dfrac{d}{dx} \dfrac{\sin x}{\cos x} = \dfrac{\cos x (d/dx) \sin x - \sin x (d/dx) \cos x}{(\cos x)^2}$

$$= \dfrac{(\cos x)^2 + (\sin x)^2}{(\cos x)^2} = \dfrac{1}{(\cos x)^2} = (\sec x)^2.$$

Similarly we have

4. $\dfrac{d}{dx} \cot x = -(\csc x)^2.$

5. $\dfrac{d}{dx} \sec x = \sec x \tan x.$

6. $\dfrac{d}{dx} \csc x = -\csc x \cot x.$

C   Derivatives of ln x and e^x: We conclude our library of derivatives of basic functions with

$$\dfrac{d}{dx} e^x = e^x$$

and

$$\dfrac{d}{dx} \ln x = \dfrac{1}{x}.$$

We may apply the Chain Rule to obtain the following particularly useful generalization of this logarithmic derivative:

$$\dfrac{d}{dx} \ln \varphi(x) = \dfrac{\varphi'(x)}{\varphi(x)}.$$

Now it is time to learn to differentiate the functions that we will commonly encounter in our work. We do so by applying the rules for sums, products, quotients, and compositions to the formulas for the derivatives of the elementary functions.

Practice is the essential tool in mastery of these ideas. Be sure to do all the **You Try It** problems in this section.

EXAMPLE 2.14
Calculate the derivative

$$\frac{d}{dx}[(\sin x + x) \cdot (x^3 - \ln x)].$$

SOLUTION

We know that $(d/dx)\sin x = \cos x$, $(d/dx)x = 1$, $(d/dx)x^3 = 3x^2$, and $(d/dx)\ln x = (1/x)$. Therefore, by the addition rule,

$$\frac{d}{dx}(\sin x + x) = \frac{d}{dx}\sin x + \frac{d}{dx}x = \cos x + 1$$

and

$$\frac{d}{dx}(x^3 - \ln x) = \frac{d}{dx}x^3 - \frac{d}{dx}\ln x = 3x^2 - \frac{1}{x}.$$

Now we may conclude the calculation by applying the product rule:

$$\frac{d}{dx}\left[(\sin x + x) \cdot (x^3 - \ln x)\right]$$
$$= \frac{d}{dx}(\sin x + x) \cdot (x^3 - \ln x) + (\sin x + x) \cdot \frac{d}{dx}(x^3 - \ln x)$$
$$= (\cos x + 1) \cdot (x^3 - \ln x) + (\sin x + x) \cdot \left(3x^2 - \frac{1}{x}\right)$$
$$= (4x^3 - 1) + \left(x^3 \cos x + 3x^2 \sin x - \frac{1}{x}\sin x\right) - (\ln x \cos x + \ln x).$$

EXAMPLE 2.15
Calculate the derivative

$$\frac{d}{dx}\left(\frac{e^x + x\sin x}{\tan x}\right).$$

SOLUTION

We know that $(d/dx)e^x = e^x$, $(d/dx)x = 1$, $(d/dx)\sin x = \cos x$, and $(d/dx)\tan x = \sec^2 x$. By the product rule,

$$\frac{d}{dx}[x \cdot \sin x] = \left(\frac{d}{dx}x\right) \cdot \sin x + x \cdot \frac{d}{dx}\sin x = 1 \cdot \sin x + x \cdot \cos x.$$

Therefore, by the quotient rule,

$$\frac{d}{dx}\left(\frac{e^x + x \sin x}{\tan x}\right) = \frac{\tan x \cdot (d/dx)(e^x + x \sin x) - (e^x + x \sin x)(d/dx)\tan x}{(\tan x)^2}$$

$$= \frac{\tan x \cdot (e^x + \sin x + x \cos x) - (e^x + x \sin x) \cdot (\sec x)^2}{(\tan x)^2}$$

$$= \frac{e^x \tan x + \tan x \sin x + x \sin x - e^x \sec^2 x - x \sin x \sec^2 x}{\tan^2 x}.$$

You Try It: Calculate the derivative $\dfrac{d}{dx}\left(\sin x \cdot \left(\cos x - \dfrac{x}{e^x + \ln x}\right)\right).$

EXAMPLE 2.16

Calculate the derivative

$$\frac{d}{dx}(\sin(x^3 - x^2)).$$

SOLUTION

This is the composition of functions, so we must apply the Chain Rule. It is essential to recognize what function will play the role of f and what function will play the role of g.

Notice that, if x is the variable, then $x^3 - x^2$ is applied first and sin applied next. So it must be that $g(x) = x^3 - x^2$ and $f(s) = \sin s$. Notice that $(d/ds)f(s) = \cos s$ and $(d/dx)g(x) = 3x^2 - 2x$. Then

$$\sin(x^3 - x^2) = f \circ g(x)$$

and

$$\frac{d}{dx}(\sin(x^3 - x^2)) = \frac{d}{dx}(f \circ g(x))$$

$$= \left[\frac{df}{ds}(g(x))\right] \cdot \frac{d}{dx}g(x)$$

$$= \cos(g(x)) \cdot (3x^2 - 2x)$$

$$= [\cos(x^3 - x^2)] \cdot (3x^2 - 2x).$$

EXAMPLE 2.17

Calculate the derivative

$$\frac{d}{dx}\ln\left(\frac{x^2}{x-2}\right).$$

SOLUTION
Let

$$h(x) = \ln\left(\frac{x^2}{x-2}\right).$$

Then

$$h = f \circ g,$$

where $f(s) = \ln s$ and $g(x) = x^2/(x-2)$. So $(d/ds)f(s) = 1/s$ and $(d/dx)g(x) = (x-2) \cdot 2x - x^2 \cdot 1/(x-2)^2 = (x^2 - 4x)/(x-2)^2$. As a result,

$$\begin{aligned}
\frac{d}{dx}h(x) &= \frac{d}{dx}(f \circ g)(x) \\
&= \left[\frac{df}{ds}(g(x))\right] \cdot \frac{d}{dx}g(x) \\
&= \frac{1}{g(x)} \cdot \frac{x^2 - 4x}{(x-2)^2} \\
&= \frac{1}{x^2/(x-2)} \cdot \frac{x^2 - 4x}{(x-2)^2} \\
&= \frac{x-4}{x(x-2)}.
\end{aligned}$$

You Try It: Perform the differentiation in the last example by first applying a rule of logarithms to simplify the function to be differentiated.

You Try It: Calculate the derivative of $\tan(e^x - x)$.

EXAMPLE 2.18

Calculate the tangent line to the graph of $f(x) = x \cdot e^{x^2}$ at the point $(1, e)$.

SOLUTION
The slope of the tangent line will be the derivative. Now

$$f'(x) = [x]' \cdot e^{x^2} + x \cdot \left[e^{x^2}\right]' = e^{x^2} + x \cdot \left[2x \cdot e^{x^2}\right].$$

In the last derivative we have of course used the Chain Rule. Thus $f'(1) = e + 2e = 3e$. Therefore the equation of the tangent line is

$$(y - e) = 3e(x - 1).$$

You Try It: Calculate the equation of the tangent line to the graph of $g(x) = \cos((x^2 - 2)/\ln x)$ at the point $(2, \cos[2/\ln 2])$.

Math Note: Calculate $(d/dx)(x^2/x)$ using the quotient rule. Of course $x^2/x = x$, and you may calculate the derivative directly. Observe that the two answers are the same. The calculation confirms the validity of the quotient rule by way of an example. Use a similar example to confirm the validity of the product rule.

2.5.1 THE DERIVATIVE OF AN INVERSE

An important formula in the calculus relates the derivative of the inverse of a function to the derivative of the function itself. The formula is

$$[f^{-1}]'(t) = \frac{1}{f'(f^{-1}(t))}. \qquad (\star)$$

We encourage you to apply the Chain Rule to the formula $f(f^{-1}(x)) = x$ to obtain a formal derivation of the formula $(\star)$.

EXAMPLE 2.19

Calculate the derivative of $g(t) = t^{1/3}$.

SOLUTION

Set $f(s) = s^3$ and apply formula $(\star)$. Then $f'(s) = 3s^2$ and $f^{-1}(t) = t^{1/3}$. With $s = f^{-1}(t)$ we then have

$$[f^{-1}]'(t) = \frac{1}{f'(f^{-1}(t))} = \frac{1}{3s^2} = \frac{1}{3 \cdot [t^{1/3}]^2} = \frac{1}{3} \cdot t^{-2/3}.$$

Formula $(\star)$ may be applied to obtain some interesting new derivatives to add to our library. We record some of them here:

   I. $\quad \dfrac{d}{dx} \arcsin x = \dfrac{1}{\sqrt{1 - x^2}}$

   II. $\quad \dfrac{d}{dx} \arccos x = -\dfrac{1}{\sqrt{1 - x^2}}$

   III. $\quad \dfrac{d}{dx} \arctan x = \dfrac{1}{1 + x^2}$

You Try It: Calculate the derivative of $f(x) = \sqrt{x}$. Calculate the derivative of $g(x) = \sqrt[k]{x}$ for any $k \in \{2, 3, 4, \dots\}$.

2.6   The Derivative as a Rate of Change

If $f(t)$ represents the position of a moving body, or the amount of a changing quantity, at time t, then the derivative $f'(t)$ (equivalently, $(d/dt)f(t)$) denotes the rate of change of position (also called *velocity*) or the rate of change of the quantity.

When $f'(t)$ represents velocity, then sometimes we calculate *another derivative*—$(f')'(t)$—and this quantity denotes the rate of change of velocity, or *acceleration*. In specialized applications, even more derivatives are sometimes used. For example, sometimes the derivative of the acceleration is called *jerk* and sometimes the derivative of jerk is called *surge*.

EXAMPLE 2.20

The position of a body moving along a linear track is given by $p(t) = 3t^2 - 5t + 7$ feet. Calculate the velocity and the acceleration at time $t = 3$ seconds.

SOLUTION

The velocity is given by

$$p'(t) = 6t - 5.$$

At time $t = 3$ we therefore find that the velocity is $p'(3) = 18 - 5 = 13$ ft/sec.

The acceleration is given by the *second derivative*:

$$p''(t) = (p')'(t) = (6t - 5)' = 6.$$

The acceleration at time $t = 3$ is therefore 6 ft/sec^2.

Math Note: As previously noted, velocity is measured in feet per second (or ft/sec). Acceleration is the rate of change of velocity with respect to time; therefore acceleration is measured in "feet per second per second" (or ft/sec^2).

EXAMPLE 2.21

A massive ball is dropped from a tower. It is known that a falling body descends (near the surface of the earth) with an acceleration of about 32 ft/sec. From this information one can determine that the equation for the position of the ball at time t is

$$p(t) = -16t^2 + v_0 t + h_0 \text{ ft.}$$

Here v_0 is the initial velocity and h_0 is the initial height of the ball in feet.[1] Also t is time measured in seconds. If the ball hits the earth after 5 seconds, then determine the height from which the ball is dropped.

SOLUTION

Observe that the velocity is

$$v(t) = p'(t) = -32t + v_0.$$

Obviously the initial velocity of a falling body is 0. Thus

$$0 = v(0) = -32 \cdot 0 + v_0.$$

---

[1] We shall say more about this equation, and this technique, in Section 3.4.

It follows that $v_0 = 0$, thus confirming our intuition that the initial velocity is 0. Thus

$$p(t) = -16t^2 + h_0.$$

Now we also know that $p(5) = 0$; that is, at time $t = 5$ the ball is at height 0. Thus

$$0 = p(5) = -16 \cdot 5^2 + h_0.$$

We may solve this equation for h_0 to determine that $h_0 = 400$.
We conclude that

$$p(t) = -16t^2 + 400.$$

Furthermore, $p(0) = 400$, so the initial height of the ball is 400 feet.

You Try It: Suppose that a massive ball falls from a height of 600 feet. After how many seconds will it strike the ground?

Exercises

1. Calculate, if possible, each of these limits. Give reasons for each step of your solution.

   (a) $\lim\limits_{x \to 0} x \cdot e^x$

   (b) $\lim\limits_{x \to 1} \dfrac{x^2 - 1}{x - 1}$

   (c) $\lim\limits_{x \to 2} (x - 2) \cdot \cot(x - 2)$

   (d) $\lim\limits_{x \to 0} x \cdot \ln x$

   (e) $\lim\limits_{t \to 3} \dfrac{t^2 - 7t + 12}{t - 3}$

   (f) $\lim\limits_{s \to 4} \dfrac{s^2 - 3s - 4}{s - 4}$

   (g) $\lim\limits_{x \to 1} \dfrac{\ln x}{x - 1}$

   (h) $\lim\limits_{x \to -3} \dfrac{x^2 - 9}{x + 3}$

2. Determine whether the given function f is continuous at the given point c. Give careful justifications for your answers.

(a) $f(x) = \dfrac{x-1}{x+1}$ $c = -1$

(b) $f(x) = \dfrac{x-1}{x+1}$ $c = 3$

(c) $f(x) = x \cdot \sin(1/x)$ $c = 0$

(d) $f(x) = x \cdot \ln x$ $c = 0$

(e) $f(x) = \begin{cases} x^2 & \text{if } x \leq 1 \\ x & \text{if } 1 < x \end{cases}$ $c = 1$

(f) $f(x) = \begin{cases} x^2 & \text{if } x \leq 1 \\ 2x & \text{if } 1 < x \end{cases}$ $c = 1$

(g) $f(x) = \begin{cases} \sin x & \text{if } x \leq \pi \\ (x - \pi) & \text{if } \pi < x \end{cases}$ $c = \pi$

(h) $f(x) = e^{\ln x + x}$ $c = 2$

3. Use the definition of derivative to calculate each of these derivatives.

(a) $f'(2)$ when $f(x) = x^2 + 4x$

(b) $f'(1)$ when $f(x) = -1/x^2$

4. Calculate each of these derivatives. Justify each step of your calculation.

(a) $\left[\dfrac{x}{x^2 + 1} \right]'$

(b) $\dfrac{d}{dx} \sin(x^2)$

(c) $\dfrac{d}{dt} t \cdot \tan(t^3 - t^2)$

(d) $\dfrac{d}{dx} \dfrac{x^2 - 1}{x^2 + 1}$

(e) $[x \cdot \ln(\sin x)]'$

(f) $\dfrac{d}{ds} e^{s(s+2)}$

(g) $\dfrac{d}{dx} e^{\sin(x^2)}$

(h) $[\ln(e^x + x)]'$

5. Imitate the example in the text to do each of these falling body problems.

    (a)   A ball is dropped from a height of 100 feet. How long will it take that ball to hit the ground?

    (b)   Suppose that the ball from part (a) is thrown straight down with an initial velocity of 10 feet per second. Then how long will it take the ball to hit the ground?

    (c)   Suppose that the ball from part (a) is thrown straight up with an initial velocity of 10 feet per second. Then how long will it take the ball to hit the ground?

6. Use the Chain Rule to perform each of these differentiations:

    (a) $\dfrac{d}{dx} \sin(\ln(\cos x))$

    (b) $\dfrac{d}{dx} e^{\sin(\cos x)}$

    (c) $\dfrac{d}{dx} \ln(e^{\sin x} + x)$

    (d) $\dfrac{d}{dx} \arcsin(x^2 + \tan x)$

    (e) $\dfrac{d}{dx} \arccos(\ln x - e^x/5)$

    (f) $\dfrac{d}{dx} \arctan(x^2 + e^x)$

7. If a car has position $p(t) = 6t^2 - 5t + 20$ feet, where t is measured in seconds, then what is the velocity of that car at time $t = 4$? What is the average velocity of that car from $t = 2$ to $t = 8$? What is the greatest velocity over the time interval $[5, 10]$?

8. In each of these problems, use the formula for the derivative of an inverse function to find $[f^{-1}]'(1)$.

    (a) $f(0) = 1,\ f'(0) = 3$

    (b) $f(3) = 1,\ f'(3) = 8$

    (c) $f(2) = 1,\ f'(2) = \pi^2$

    (d) $f(1) = 1,\ f'(1) = 40$

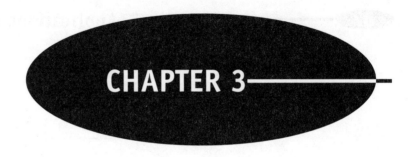

Applications of the Derivative

3.1   Graphing of Functions

We know that the value of the derivative of a function f at a point x represents the slope of the tangent line to the graph of f at the point $(x, f(x))$. If that slope is positive, then the tangent line rises as x increases from left to right, hence so does the curve (we say that the function is *increasing*). If instead the slope of the tangent line is negative, then the tangent line falls as x increases from left to right, hence so does the curve (we say that the function is *decreasing*). We summarize:

On an interval where $f' > 0$ the graph of f goes uphill.

On an interval where $f' < 0$ the graph of f goes downhill.

See Fig. 3.1.

With some additional thought, we can also get useful information from the second derivative. If $f'' = (f')' > 0$ at a point, then f' is increasing. Hence the slope of the tangent line is getting ever greater (the graph is *concave up*). The picture must be as in Fig. 3.2(*a*) or 3.2(*b*). If instead $f'' = (f')' < 0$ at a point then f' is decreasing. Hence the slope of the tangent line is getting ever less (the graph is *concave down*). The picture must be as in Fig. 3.3(*a*) or 3.3(*b*).

Using information about the first and second derivatives, we can render rather accurate graphs of functions. We now illustrate with some examples.

EXAMPLE 3.1

Sketch the graph of $f(x) = x^2$.

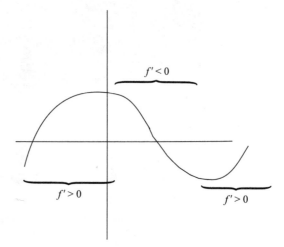

Fig. 3.1

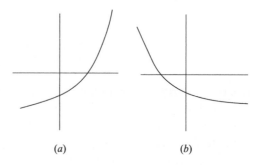

(a)   (b)

Fig. 3.2

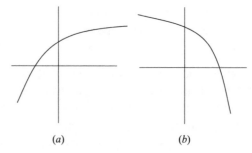

(a)   (b)

Fig. 3.3

SOLUTION

Of course this is a simple and familiar function, and you know that its graph is a parabola. But it is satisfying to see calculus confirm the shape of the graph. Let us see how this works.

First observe that $f'(x) = 2x$. We see that $f' < 0$ when $x < 0$ and $f' > 0$ when $x > 0$. So the graph is decreasing on the negative real axis and the graph is increasing on the positive real axis.

Next observe that $f''(x) = 2$. Thus $f'' > 0$ at all points. Thus the graph is concave up everywhere.

Finally note that the graph passes through the origin.

We summarize this information in the graph shown in Fig. 3.4.

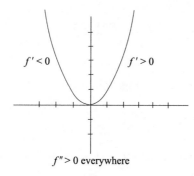

$f' < 0$ $f' > 0$

$f'' > 0$ everywhere

Fig. 3.4

EXAMPLE 3.2

Sketch the graph of $f(x) = x^3$.

SOLUTION

First observe that $f'(x) = 3x^2$. Thus $f' \geq 0$ everywhere. The function is always increasing.

Second observe that $f''(x) = 6x$. Thus $f''(x) < 0$ when $x < 0$ and $f''(x) > 0$ when $x > 0$. So the graph is concave down on the negative real axis and concave up on the positive real axis.

Finally note that the graph passes through the origin.

We summarize our findings in the graph shown in Fig. 3.5.

You Try It: Use calculus to aid you in sketching the graph of $f(x) = x^3 + x$.

EXAMPLE 3.3

Sketch the graph of $g(x) = x + \sin x$.

SOLUTION

We see that $g'(x) = 1 + \cos x$. Since $-1 \leq \cos x \leq 1$, it follows that $g'(x) \geq 0$. Hence the graph of g is always increasing.

Now $g''(x) = -\sin x$. This function is positive sometimes and negative sometimes. In fact

$$-\sin x \text{ is positive when } k\pi < x < (k+1)\pi, \ k \text{ odd}$$

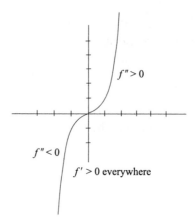

Fig. 3.5

and

$$-\sin x \text{ is negative when } k\pi < x < (k+1)\pi, \; k \text{ even.}$$

So the graph alternates being concave down and concave up. Of course it also passes through the origin. We amalgamate all our information in the graph shown in Fig. 3.6.

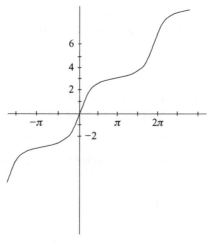

Fig. 3.6

EXAMPLE 3.4

Sketch the graph of $h(x) = x/(x+1)$.

SOLUTION
First note that the function is undefined at $x = -1$.

We calculate that $h'(x) = 1/((x + 1)^2)$. Thus the graph is everywhere increasing (except at $x = -1$).

We also calculate that $h''(x) = -2/((x + 1)^3)$. Hence $h'' > 0$ and the graph is concave up when $x < -1$. Likewise $h'' < 0$ and the graph is concave down when $x > -1$.

Finally, as x tends to -1 from the left we notice that h tends to $+\infty$ and as x tends to -1 from the right we see that h tends to $-\infty$.

Putting all this information together, we obtain the graph shown in Fig. 3.7.

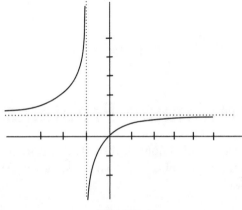

Fig. 3.7

You Try It: Sketch the graph of the function $k(x) = x \cdot \sqrt{x + 1}$.

EXAMPLE 3.5
Sketch the graph of $k(x) = x^3 + 3x^2 - 9x + 6$.

SOLUTION
We notice that $k'(x) = 3x^2 + 6x - 9 = 3(x - 1)(x + 3)$. So the *sign* of k' changes at $x = 1$ and $x = -3$. We conclude that

k' is positive when $x < -3$;

k' is negative when $-3 < x < 1$;

k' is positive when $x > 3$.

Finally, $k''(x) = 6x + 6$. Thus the graph is concave down when $x < -1$ and the graph is concave up when $x > -1$.

Putting all this information together, and using the facts that $k(x) \to -\infty$ when $x \to -\infty$ and $k(x) \to +\infty$ when $x \to +\infty$, we obtain the graph shown in Fig. 3.8.

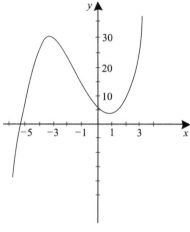

Fig. 3.8

3.2 Maximum/Minimum Problems

One of the great classical applications of the calculus is to determine the maxima and minima of functions. Look at Fig. 3.9. It shows some (local) maxima and (local) minima of the function f.

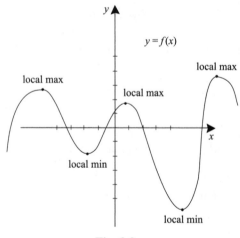

Fig. 3.9

Notice that a maximum has the characterizing property that it looks like a hump: the function is increasing to the left of the hump and decreasing to the right of the hump. The derivative *at the hump* is 0: the function neither increases nor decreases at a local maximum. This is sometimes called *Fermat's test*. Also, we see that the graph is concave down at a local maximum.

It is common to refer to the points where the derivative vanishes as *critical points*. In some contexts, we will designate the endpoints of the domain of our function to be critical points as well.

Now look at a local minimum. Notice that a minimum has the characterizing property that it looks like a valley: the function is decreasing to the left of the valley and increasing to the right of the valley. The derivative *at the valley* is 0: the function neither increases nor decreases at a local minimum. This is another manifestation of Fermat's test. Also, we see that the graph is concave up at a local minimum.

Let us now apply these mathematical ideas to some concrete examples.

EXAMPLE 3.6

Find all local maxima and minima of the function $k(x) = x^3 - 3x^2 - 24x + 5$.

SOLUTION

We begin by calculating the first derivative:

$$k'(x) = 3x^2 - 6x - 24 = 3(x + 2)(x - 4).$$

We notice that k' vanishes only when $x = -2$ or $x = 4$. These are the only candidates for local maxima or minima. The second derivative is $k''(x) = 6x - 6$. Now $k''(4) = 18 > 0$, so $x = 4$ is the location of a local minimum. Also $k''(-2) = -18 < 0$, so $x = -2$ is the location of a local maximum. A glance at the graph of this function, as depicted in Fig. 3.10, confirms our calculations.

EXAMPLE 3.7

Find all local maxima and minima of the function $g(x) = x + \sin x$.

SOLUTION

First we calculate that

$$g'(x) = 1 + \cos x.$$

Thus g' vanishes at the points $(2k + 1)\pi$ for $k = \ldots, -2, -1, 0, 1, 2, \ldots$. Now $g''(x) = \sin x$. And $g''((2k + 1)\pi) = 0$. Thus the second derivative test is inconclusive. Let us instead look at the first derivative. We notice that it is always ≥ 0. But, as we have already noticed, the first derivative changes sign at a local maximum or minimum. We conclude that none of the points $(2k + 1)\pi$ is either a maximum nor a minimum. The graph in Fig. 3.11 confirms this calculation.

You Try It: Find all local maxima and minima of the function $g(x) = 2x^3 - 15x^2 + 24x + 6$.

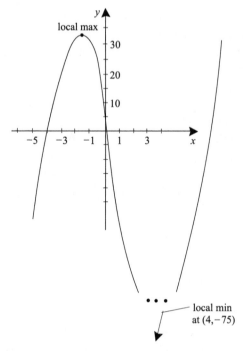

Fig. 3.10

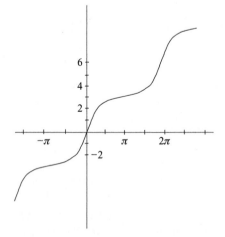

Fig. 3.11

EXAMPLE 3.8

A box is to be made from a sheet of cardboard that measures 12″ × 12″.
The construction will be achieved by cutting a square from each corner of

the sheet and then folding up the sides (see Fig. 3.12). What is the box of greatest volume that can be constructed in this fashion?

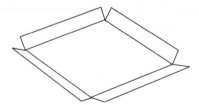

Fig. 3.12

SOLUTION

It is important in a problem of this kind to introduce a variable. Let x be the side length of the squares that are to be cut from the sheet of cardboard. Then the side length of the resulting box will be $12 - 2x$ (see Fig. 3.13). Also the *height* of the box will be x. As a result, the volume of the box will be

$$V(x) = x \cdot (12 - 2x) \cdot (12 - 2x) = 144x - 48x^2 + 4x^3.$$

Our job is to maximize this function V.

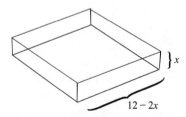

Fig. 3.13

Now $V'(x) = 144 - 96x + 12x^2$. We may solve the quadratic equation

$$144 - 96x + 12x^2 = 0$$

to find the critical points for this problem. Using the quadratic formula, we find that $x = 2$ and $x = 6$ are the critical points. Now $V''(x) = -96 + 24x$. Since $V''(2) = -48 < 0$, we conclude that $x = 2$ is a local maximum for the problem. In fact we can sketch a graph of $V(x)$ using ideas from calculus and see that $x = 2$ is a global maximum.

We conclude that if squares of side $2''$ are cut from the sheet of cardboard then a box of maximum volume will result.

Observe in passing that if squares of side $6''$ are cut from the sheet then (there will be no cardboard left!) the resulting box will have zero volume. This value for x gives a minimum for the problem.

EXAMPLE 3.9

A rectangular garden is to be constructed against the side of a garage. The gardener has 100 feet of fencing, and will construct a three-sided fence; the side of the garage will form the fourth side. What dimensions will give the garden of greatest area?

SOLUTION

Look at Fig. 3.14. Let x denote the side of the garden that is perpendicular to the side of the garage. Then the resulting garden has width x feet and length $100 - 2x$ feet. The area of the garden is

$$A(x) = x \cdot (100 - 2x) = 100x - 2x^2.$$

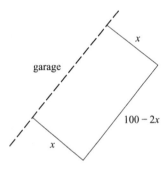

Fig. 3.14

We calculate $A'(x) = 100 - 4x$ and find that the only critical point for the problem is $x = 25$. Since $A''(x) = -4$ for all x, we determine that $x = 25$ is a local maximum. By inspection, we see that the graph of A is a downward-opening parabola. So $x = 25$ must also be the global maximum that we seek. The optimal dimensions for the garden are

$$\text{width} = 25 \, \text{ft.} \qquad \text{length} = 50 \, \text{ft.}$$

You Try It: Find the right circular cylinder of greatest volume that can be contained in a sphere of radius 1.

EXAMPLE 3.10

The sum of two positive numbers is 60. How can we choose them so as to maximize their product?

SOLUTION

Let x be one of the two numbers. Then the other is $60 - x$. Their product is

$$P(x) = x \cdot (60 - x) = 60x - x^2.$$

Thus P is the quantity that we wish to maximize. Calculating the derivative, we find that

$$P'(x) = 60 - 2x.$$

Thus the only critical point for the problem is $x = 30$. Since $P''(x) \equiv -2$, we find that $x = 30$ is a local maximum. Since the graph of P is a downward-opening parabola, we can in fact be sure that $x = 30$ is a global maximum.

We conclude that the two numbers that add to 60 and maximize the product are 30 and 30.

You Try It: A rectangular box is to be placed in the first quadrant $\{(x, y) : x \geq 0, y \geq 0\}$ in such a way that one side lies on the positive x-axis and one side lies on the positive y-axis. The box is to lie below the line $y = -2x + 5$. Give the dimensions of such a box having greatest possible area.

3.3   Related Rates

If a tree is growing in a forest, then both its height and its radius will be increasing. These two growths will depend in turn on (i) the amount of sunlight that hits the tree, (ii) the amount of nutrients in the soil, (iii) the proximity of other trees. We may wish to study the relationship among these various parameters. For example, if we know that the amount of sunlight and nutrients are increasing at a certain rate then we may wish to know how that affects the rate of change of the radius. This consideration gives rise to *related rates* problems.

EXAMPLE 3.11

A toy balloon is in the shape of a sphere. It is being inflated at the rate of 20 in.^{3}/min. At the moment that the sphere has volume 64 cubic inches, what is the rate of change of the radius?

SOLUTION

We know that volume and radius of a sphere are related by the formula

$$V = \frac{4\pi}{3}r^3. \qquad\qquad (*)$$

The free variable in this problem is time, so we differentiate equation $(*)$ with respect to time t. It is important that we keep the chain rule in mind as we do so.[1] The result is

$$\frac{dV}{dt} = \frac{4\pi}{3} \cdot 3r^2 \cdot \frac{dr}{dt}. \qquad\qquad (**)$$

─────────────
[1] The point is that we are *not* differentiating with respect to r.

Now we are given that $dV/dt = 20$. Our question is posed at the moment that $V = 64$. But, according to (∗), this means that $r = \sqrt[3]{48/\pi}$. Substituting these values into equation (∗∗) yields

$$20 = \frac{4\pi}{3} \cdot 3\left[\sqrt[3]{48/\pi}\,\right]^2 \cdot \frac{dr}{dt}.$$

Solving for dr/dt yields

$$\frac{dr}{dt} = \frac{5}{48^{2/3} \cdot \pi^{1/3}}.$$

EXAMPLE 3.12

A 13-foot ladder leans against a wall (Fig. 3.15). The foot of the ladder begins to slide away from the wall at the rate of 1 foot per minute. When the foot is 5 feet from the wall, at what rate is the top of the ladder falling?

Fig. 3.15

SOLUTION

Let $h(t)$ be the height of the ladder at time t and $b(t)$ be the distance of the base of the ladder to the wall at time t. Then the Pythagorean theorem tells us that

$$h(t)^2 + b(t)^2 = 13^2.$$

We may differentiate both sides of this equation with respect to the variable t (which is time in minutes) to obtain

$$2 \cdot h(t) \cdot h'(t) + 2 \cdot b(t) \cdot b'(t) = 0.$$

Solving for $h'(t)$ yields

$$h'(t) = -\frac{b(t) \cdot b'(t)}{h(t)}.$$

At the instant the problem is posed, $b(t) = 5$, $h(t) = 12$ (by the Pythagorean theorem), and $b'(t) = 1$. Substituting these values into the equation yields

$$h'(t) = -\frac{5 \cdot 1}{12} = -\frac{5}{12} \text{ ft/min.}$$

Observe that the answer is negative, which is appropriate since the top of the ladder is falling.

You Try It: Suppose that a square sheet of aluminum is placed in the hot sun. It begins to expand very slowly so that its diagonal is increasing at the rate of 1 millimeter per minute. At the moment that the diagonal is 100 millimeters, at what rate is the area increasing?

EXAMPLE 3.13

A sponge is in the shape of a right circular cone (Fig. 3.16). As it soaks up water, it grows in size. At a certain moment, the height equals 6 inches, and is increasing at the rate of 0.3 inches per second. At that same moment, the radius is 4 inches, and is increasing at the rate of 0.2 inches per second. How is the volume changing at that time?

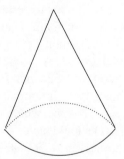

Fig. 3.16

SOLUTION

We know that the volume V of a right circular cone is related to the height h and the radius r by the formula

$$V = \frac{1}{3}\pi r^2 h.$$

Differentiating both sides with respect to the variable t (for time in seconds) yields

$$\frac{dV}{dt} = \frac{1}{3}\pi \left[2r\frac{dr}{dt}h + r^2\frac{dh}{dt} \right].$$

Substituting the values for r, dr/dt, h, and dh/dt into the right-hand side yields

$$\frac{dV}{dt} = \frac{1}{3}\pi \left[2 \cdot 4 \cdot (0.2) \cdot 6 + 4^2 \cdot (0.3) \right] = \frac{1}{3}\pi [9.6 + 4.8] = \frac{24\pi}{5}.$$

You Try It: In the heat of the sun, a sheet of aluminum in the shape of an equilateral triangle is expanding so that its side length increases by 1 millimeter per hour. When the side length is 100 millimeters, how is the area increasing?

3.4   Falling Bodies

It is known that, near the surface of the earth, a body falls with acceleration (due to gravity) of about 32 ft/sec^2. If we let $h(t)$ be the height of the object at time t (measured in seconds), then our information is that

$$h''(t) = -32.$$

Observe the minus sign to indicate that height is decreasing.

Now we will do some organized guessing. What could h' be? It is some function whose derivative is the constant -32. Our experience indicates that polynomials decrease in degree when we differentiate them. That is, the degree goes from 5 to 4, or from 3 to 2. Since, h'' is a polynomial of degree 0, we therefore determine that h' will be a polynomial of degree 1. A moment's thought then suggests that $h'(t) = -32t$. This works! If $h'(t) = -32t$ then $h''(t) = [h'(t)]' = -32$. In fact we can do a bit better. Since constants differentiate to zero, our best guess of what the velocity should be is $h'(t) = -32t + v_0$, where v_0 is an undetermined constant.

Now let us guess what form $h(t)$ should have. We can learn from our experience in the last paragraph. The "antiderivative" of $-32t$ (a polynomial of degree 1) should be a polynomial of degree 2. After a little fiddling, we guess $-16t^2$. And this works. The antiderivative of v_0 (a polynomial of degree 0) should be a polynomial of degree 1. After a little fiddling, we guess $v_0 t$. And this works. Taking all this information together, we find that the "antiderivative" of $h'(t) = -32t + v_0$ is

$$h(t) = -16t^2 + v_0 t + h_0. \tag{$\dagger$}$$

Notice that we have once again thrown in an additive constant h_0. This does no harm:

$$h'(t) = [-16t^2]' + [v_0 t]' + [h_0]' = -32t + v_0,$$

just as we wish. And, to repeat what we have already confirmed,

$$h''(t) = [h'(t)]' = [-32t]' + [v_0]' = -32.$$

We now have a general formula (namely (†)) for the position of a falling body at time t. [Recall that we were first introduced to this formula in Section 2.6.] See Fig. 3.17.

Fig. 3.17

Before doing some examples, we observe that a falling body will have initial velocity 0. Thus

$$0 = h'(0) = -32 \cdot 0 + v_0.$$

Hence, for a falling body, $v_0 = 0$. In some problems we may give the body an initial push, and then v_0 will not be zero.

EXAMPLE 3.14

Suppose that a falling body hits the ground with velocity -100 ft/sec. What was the initial height of the body?

SOLUTION

With notation as developed above, we know that velocity is given by

$$h'(t) = -32t + 0.$$

We have taken v_0 to be 0 because the body is a falling body; it had no initial push. If T is the time at which the body hits the ground, then we know that

$$-100 = h'(T) = -32 \cdot T.$$

As a result, $T = 25/8$ sec.

When the body hits the ground, its height is 0. Thus we know that

$$0 = h(T) = h(25/8) = -16 \cdot (25/8)^2 + h_0.$$

We may solve for h_0 to obtain

$$h_0 = \frac{625}{4}.$$

Thus all the information about our falling body is given by

$$h(t) = -16t^2 + \frac{625}{4}.$$

At time $t = 0$ we have

$$h(0) = \frac{625}{4}.$$

Thus the initial height of the falling body is $625/4 \, \text{ft} = 156.25 \, \text{ft}$.

Notice that, in the process of solving the last example, and in the discussion preceding it, we learned that h_0 represents the initial height of the falling body and v_0 represents the initial velocity of the falling body. This information will be useful in the other examples that we examine.

EXAMPLE 3.15

A body is thrown straight down with an initial velocity of 10 feet per second. It strikes the ground in 12 seconds. What was the initial height?

SOLUTION

We know that $v_0 = -10$ and that $h(12) = 0$. This is the information that we must exploit in solving the problem. Now

$$h(t) = -16t^2 - 10t + h_0.$$

Thus

$$0 = h(12) = -16 \cdot 12^2 - 10 \cdot 12 + h_0.$$

We may solve for h_0 to obtain

$$h_0 = 2424 \, \text{ft}.$$

The initial height is 2424 feet.

You Try It: A body is thrown straight up with initial velocity 5 feet per second from a height of 40 feet. After how many seconds will it hit the ground? What will be its maximum height?

EXAMPLE 3.16

A body is launched straight up from height 100 feet with some initial velocity. It hits the ground after 10 seconds. What was that initial velocity?

SOLUTION

We are given that $h_0 = 100$. Thus

$$h(t) = -16t^2 + v_0 t + 100.$$

Our job is to find v_0. We also know that

$$0 = h(10) = -16 \cdot 10^2 + v_0 \cdot 10 + 100.$$

We solve this equation to find that $v_0 = 150\,\text{ft/sec}$.

You Try It: On a certain planet, bodies fall with an acceleration due to gravity of $10\,\text{ft/sec}^2$. A certain body is thrown down with an initial velocity of 5 feet per second, and hits the surface 12 seconds later. From what height was it launched?

Exercises

1. Sketch the graph of $f(x) = x/[x^2 + 3]$, indicating all local maxima and minima together with concavity properties.
2. What is the right circular cylinder of greatest volume that can be inscribed upright in a right circular cone of radius 3 and height 6?
3. An air mattress (in the shape of a rectangular parallelepiped) is being inflated in such a way that, at a given moment, its length is increasing by 1 inch per minute, its width is decreasing by 0.5 inches per minute, and its height is increasing by 0.3 inches per minute. At that moment its dimensions are $\ell = 100''$, $w = 60''$, and $h = 15''$. How is its volume changing at that time?
4. A certain body is thrown straight down at an initial velocity of $15\,\text{ft/sec}$. It strikes the ground in 5 seconds. What is its initial height?
5. Because of viral infection, the shape of a certain cone-shaped cell is changing. The height is increasing at the rate of 3 microns per minute. For metabolic reasons, the volume remains constantly equal to 20 cubic microns. At the moment that the radius is 5 microns, what is the rate of change of the radius of the cell?
6. A silo is to hold 10,000 cubic feet of grain. The silo will be cylindrical in shape and have a flat top. The floor of the silo will be the earth. What dimensions of the silo will use the least material for construction?
7. Sketch the graph of the function $g(x) = x \cdot \sin x$. Show maxima and minima.

8. A body is launched straight down at a velocity of $5\,\text{ft/sec}$ from height 400 feet. How long will it take this body to reach the ground?

9. Sketch the graph of the function $h(x) = x/(x^2 - 1)$. Exhibit maxima, minima, and concavity.

10. A punctured balloon, in the shape of a sphere, is losing air at the rate of $2\,\text{in.}^3/\text{sec}$. At the moment that the balloon has volume 36π cubic inches, how is the radius changing?

11. A ten-pound stone and a twenty-pound stone are each dropped from height 100 feet at the same moment. Which will strike the ground first?

12. A man wants to determine how far below the surface of the earth is the water in a well. How can he use the theory of falling bodies to do so?

13. A rectangle is to be placed in the first quadrant, with one side on the x-axis and one side on the y-axis, so that the rectangle lies below the line $3x + 5y = 15$. What dimensions of the rectangle will give greatest area?

14. A rectangular box with square base is to be constructed to hold 100 cubic inches. The material for the base and the top costs 10 cents per square inch and the material for the sides costs 20 cents per square inch. What dimensions will give the most economical box?

15. Sketch the graph of the function $f(x) = [x^2 - 1]/[x^2 + 1]$. Exhibit maxima, minima, and concavity.

16. On the planet Zork, the acceleration due to gravity of a falling body near the surface of the planet is $20\,\text{ft/sec}$. A body is dropped from height 100 feet. How long will it take that body to hit the surface of Zork?

CHAPTER 4

The Integral

4.0 Introduction

Many processes, both in mathematics and in nature, involve addition. You are familiar with the discrete process of addition, in which you add finitely many numbers to obtain a sum or aggregate. But there are important instances in which we wish to add infinitely many terms. One important example is in the calculation of area—especially the area of an unusual (non-rectilinear) shape. A standard strategy is to approximate the desired area by the sum of small, thin rectangular regions (whose areas are easy to calculate). A second example is the calculation of work, in which we think of the work performed over an interval or curve as the aggregate of small increments of work performed over very short intervals. We need a mathematical formalism for making such summation processes natural and comfortable. Thus we will develop the concept of the integral.

4.1 Antiderivatives and Indefinite Integrals

4.1.1 THE CONCEPT OF ANTIDERIVATIVE

Let f be a given function. We have already seen in the theory of falling bodies (Section 3.4) that it can be useful to find a function F such that $F' = f$. We call such a function F an *antiderivative* of f. In fact we often want to find the most general function F, or a *family of functions*, whose derivative equals f. We can sometimes achieve this goal by a process of organized guessing.

Suppose that $f(x) = \cos x$. If we want to guess an antiderivative, then we are certainly *not* going to try a polynomial. For if we differentiate a polynomial then we get another polynomial. So that will not do the job. For similar reasons we are not going to guess a logarithm or an exponential. In fact the way that we get a trigonometric function through differentiation is by differentiating another trigonometric function. What trigonometric function, when differentiated, gives $\cos x$? There are only six functions to try, and a moment's thought reveals that $F(x) = \sin x$ does the trick. In fact an even better answer is $F(x) = \sin x + C$. The constant differentiates to 0, so $F'(x) = f(x) = \cos x$. We have seen in our study of falling bodies that the additive constant gives us a certain amount of flexibility in solving problems.

Now suppose that $f(x) = x^2$. We have already noted that the way to get a polynomial through differentiation is to differentiate another polynomial. Since differentiation reduces the degree of the polynomial by 1, it is natural to guess that the F we seek is a polynomial of degree 3. What about $F(x) = x^3$? We calculate that $F'(x) = 3x^2$. That does not quite work. We seek x^2 for our derivative, but we got $3x^2$. This result suggests adjusting our guess. We instead try $F(x) = x^3/3$. Then, indeed, $F'(x) = 3x^2/3 = x^2$, as desired. We will write $F(x) = x^3/3 + C$ for our antiderivative.

More generally, suppose that $f(x) = ax^3 + bx^2 + cx + d$. Using the reasoning in the last paragraph, we may find fairly easily that $F(x) = ax^4/4 + bx^3/3 + cx^2/2 + dx + e$. Notice that, once again, we have thrown in an additive constant.

You Try It: Find a family of antiderivatives for the function $f(x) = \sin 2x - x^4 + e^x$.

4.1.2 THE INDEFINITE INTEGRAL

In practice, it is useful to have a compact notation for the antiderivative. What we do, instead of saying that "the antiderivative of $f(x)$ is $F(x) + C$," is to write

$$\int f(x)\, dx = F(x) + C.$$

So, for example,

$$\int \cos x\, dx = \sin x + C$$

and

$$\int x^3 + x\, dx = \frac{x^4}{4} + \frac{x^2}{2} + C$$

and

$$\int e^{2x}\, dx = \frac{e^{2x}}{2} + C.$$

The symbol $\int$ is called an *integral sign* (the symbol is in fact an elongated "S") and the symbol "dx" plays a traditional role to remind us what the variable is. We call an expression like

$$\int f(x)\, dx$$

an *indefinite integral*. The name comes from the fact that later on we will have a notion of "definite integral" that specifies what value C will take—so it is more definite in the answer that it gives.

EXAMPLE 4.1

Calculate

$$\int \sin(3x + 1)\, dx.$$

SOLUTION

We know that we must guess a trigonometric function. Running through the choices, cosine seems like the best candidate. The derivative of $\cos x$ is $-\sin x$. So we immediately see that $-\cos x$ is a better guess—its derivative is $\sin x$. But then we adjust our guess to $F(x) = -\cos(3x + 1)$ to take into account the form of the argument. This almost works: we may calculate that $F'(x) = 3\sin(3x + 1)$. We determine that we must adjust by a factor of $1/3$. Now we can record our final answer as

$$\int \sin(3x + 1)\, dx = -\frac{1}{3}\cos(3x + 1) + C.$$

We invite the reader to verify that the derivative of the answer on the right-hand side gives $\sin(3x + 1)$.

EXAMPLE 4.2

Calculate

$$\int \frac{x}{x^2 + 3}\, dx.$$

SOLUTION

We notice that the numerator of the fraction is nearly the derivative of the denominator. Put in other words, if we were asked to integrate

$$\frac{2x}{x^2 + 3}$$

then we would see that we are integrating an expression of the form

$$\frac{\varphi'(x)}{\varphi(x)}$$

(which we in fact encountered among our differentiation rules in Section 2.5). As we know, expressions like this arise from differentiating $\log \varphi(x)$.

Returning to the original problem, we pose our initial guess as $\log[x^2 + 3]$. Differentiation of this expression gives the answer $2x/[x^2 + 3]$. This is close to what we want, but we must adjust by a factor of $1/2$. We write our final answer as

$$\int \frac{x}{x^2 + 3}\, dx = \frac{1}{2}\log[x^2 + 3] + C.$$

You Try It: Calculate the indefinite integral

$$\int xe^{3x^2+5}\, dx.$$

EXAMPLE 4.3

Calculate the indefinite integral

$$\int (x^3 + x^2 + 1)^{50} \cdot (6x^2 + 4x)\, dx.$$

SOLUTION

We observe that the expression $6x^2 + 4x$ is nearly the derivative of $x^3 + x^2 + 1$. In fact if we set $\varphi(x) = x^3 + x^2 + 1$ then the integrand (the quantity that we are asked to integrate) is

$$[\varphi(x)]^{50} \cdot 2\varphi'(x).$$

It is natural to guess as our antiderivative $[\varphi(x)]^{51}$. Checking our work, we find that

$$([\varphi(x)]^{51})' = 51[\varphi(x)]^{50} \cdot \varphi'(x).$$

We see that the answer obtained is quite close to the answer we seek; it is off by a numerical factor of $2/51$. With this knowledge, we write our final answer as

$$\int (x^3 + x^2 + 1)^{50} \cdot (6x^2 + 4x)\, dx = \frac{2}{51} \cdot [x^3 + x^2 + 1]^{51} + C.$$

You Try It: Calculate the indefinite integral

$$\int \frac{x^2}{x^3 + 5}\, dx.$$

4.2   Area

Consider the curve shown in Fig. 4.1. The curve is the graph of $y = f(x)$. We set for ourselves the task of calculating the area A that is (i) *under* the curve, (ii) *above* the x-axis, and (iii) *between* $x = a$ and $x = b$. Refer to Fig. 4.2 to see the geometric region we are considering.

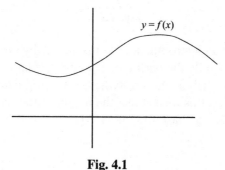

Fig. 4.1

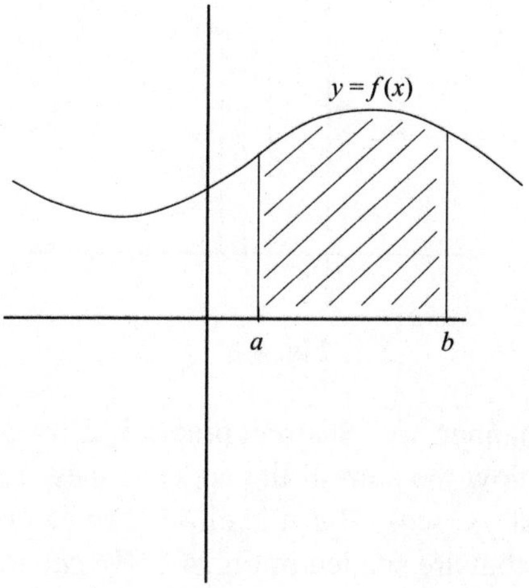

Fig. 4.2

We take it for granted that the area of a rectangle of length ℓ and width w is $\ell \times w$. Now our strategy is to divide the base interval $[a, b]$ into equal subintervals. Fix an integer $k > 0$. We designate the points

$$\mathcal{P} = \{x_0, x_1, x_2, \ldots, x_k\},$$

with $x_0 = a$ and $x_k = b$. We require that $|x_j - x_{j-1}| = |b - a|/k \equiv \Delta x$ for $j = 1, \ldots, k$. In other words, the points $x_0, x_1, \ldots, x_k$ are equally spaced. We call the set $\mathcal{P}$ a *partition*. Sometimes, to be more specific, we call it a *uniform partition* (to indicate that all the subintervals have the same length). Refer to Fig. 4.3.

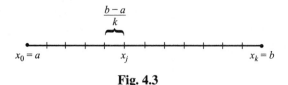

Fig. 4.3

The idea is to build an approximation to the area A by erecting rectangles over the segments determined by the partition. The first rectangle R_1 will have as base the interval $[x_0, x_1]$ and height chosen so that the rectangle touches the curve at its upper right hand corner; this means that the height of the rectangle is $f(x_1)$. The second rectangle R_2 has as base the interval $[x_1, x_2]$ and height $f(x_2)$. Refer to Fig. 4.4.

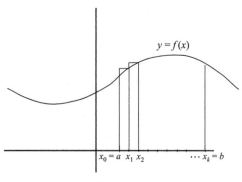

Fig. 4.4

Continuing in this manner, we construct precisely k rectangles, $R_1, R_2, \ldots, R_k$, as shown in Fig. 4.5. Now the sum of the areas of these rectangles is *not* exactly equal to the area A that we seek. But it is close. The error is the sum of the little semi-triangular pieces that are shaded in Fig. 4.6. *We can make that error as small as we please by making the partition finer.* Figure 4.7 illustrates this idea.

Let us denote by $\mathcal{R}(f, \mathcal{P})$ the sum of the areas of the rectangles that we created from the partition $\mathcal{P}$. This is called a *Riemann sum*. Thus

$$\mathcal{R}(f, \mathcal{P}) = \sum_{j=1}^{k} f(x_j) \cdot \Delta x \equiv f(x_1) \cdot \Delta x + f(x_2) \cdot \Delta x + \cdots + f(x_k) \cdot \Delta x.$$

Here the symbol $\sum_{j=1}^{k}$ denotes the sum of the expression to its right for each of the instances $j = 1$ to $j = k$.

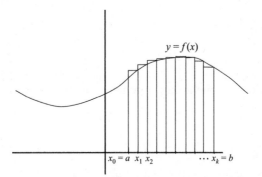

Fig. 4.5

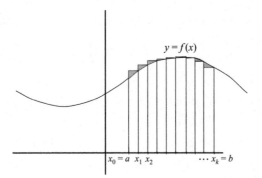

Fig. 4.6

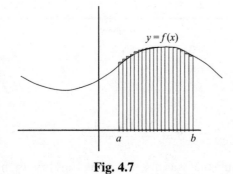

Fig. 4.7

The reasoning just presented suggests that the *true area A* is given by

$$\lim_{k \to \infty} \mathcal{R}(f, \mathcal{P}).$$

We call this limit *the integral of f from x = a to x = b* and we write it as

$$\int_a^b f(x)\, dx.$$

Thus we have learned that

$$\text{area } A = \int_a^b f(x)\,dx.$$

It is well to take a moment and comment on the integral notation. First, the integral sign

$$\int$$

is an elongated "S," coming from "summation." The dx is an historical artifact, coming partly from traditional methods of developing the integral, and partly from a need to know explicitly what the variable is. The numbers a and b are called the *limits of integration*—the number a is the lower limit and b is the upper limit. The function f is called the *integrand*.

Before we can present a detailed example, we need to record some important information about sums:

I. We need to calculate the sum $S = 1 + 2 + \cdots + N = \sum_{j=1}^{N} j$. To achieve this goal, we write

$$S = 1 + 2 \qquad + \cdots + (N-1) + N$$
$$S = N + (N-1) + \cdots + 2 \qquad + 1$$

Adding each column, we obtain

$$2S = \underbrace{(N+1) + (N+1) + \cdots + (N+1) + (N+1)}_{N \text{ times}}.$$

Thus

$$2S = N \cdot (N+1)$$

or

$$S = \frac{N \cdot (N+1)}{2}.$$

This is a famous formula that was discovered by Carl Friedrich Gauss (1777–1855) when he was a child.

II. The sum $S = 1^2 + 2^2 + \cdots + N^2 = \sum_{j=1}^{n} j^2$ is given by

$$S = \frac{2N^3 + 3N^2 + N}{6}.$$

We shall not provide the details of the proof of this formula, but refer the interested reader to [SCH2].

For our first example, we calculate the area under a parabola.

EXAMPLE 4.4

Calculate the area under the curve $y = x^2$, above the x-axis, and between $x = 0$ and $x = 2$.

SOLUTION

Refer to Fig. 4.8 as we reason along. Let $f(x) = x^2$.

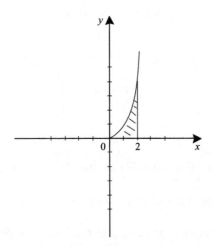

Fig. 4.8

Consider the partition $\mathcal{P}$ of the interval $[0, 2]$ consisting of $k + 1$ points $x_0, x_1, \ldots, x_k$. The corresponding Riemann sum is

$$R(f, \mathcal{P}) = \sum_{j=1}^{k} f(x_j) \cdot \Delta x.$$

Of course

$$\Delta x = \frac{2 - 0}{k} = \frac{2}{k}$$

and

$$x_j = j \cdot \frac{2}{k}.$$

In addition,

$$f(x_j) = \left(j \cdot \frac{2}{k} \right)^2 = \frac{4 j^2}{k^2}.$$

As a result, the Riemann sum for the partition $\mathcal{P}$ is

$$\mathcal{R}(f, \mathcal{P}) = \sum_{j=1}^{k} \frac{4j^2}{k^2} \cdot \frac{2}{k}$$

$$= \sum_{j=1}^{k} \frac{8j^2}{k^3} = \frac{8}{k^3} \sum_{j=1}^{k} j^2.$$

Now formula **II** above enables us to calculate the last sum explicitly. The result is that

$$\mathcal{R}(f, \mathcal{P}) = \frac{8}{k^3} \cdot \frac{2k^3 + 3k^2 + k}{6}$$

$$= \frac{8}{3} + \frac{4}{k} + \frac{4}{3k^2}.$$

In sum,

$$\int_0^2 x^2 \, dx = \lim_{k \to \infty} \mathcal{R}(f, \mathcal{P}) = \lim_{k \to \infty} \left[\frac{8}{3} + \frac{4}{k} + \frac{4}{3k^2} \right] = \frac{8}{3}.$$

We conclude that the desired area is 8/3.

You Try It: Use the method presented in the last example to calculate the area under the graph of $y = 2x$ and above the x-axis, between $x = 1$ and $x = 2$. You should obtain the answer 3, which of course can also be determined by elementary considerations—without taking limits.

The most important idea in all of calculus is that it is possible to calculate an integral without calculating Riemann sums and passing to the limit. This is the Fundamental Theorem of Calculus, due to Leibniz and Newton. We now state the theorem, illustrate it with examples, and then briefly discuss why it is true.

Theorem 4.1 (Fundamental Theorem of Calculus)
Let f be a continuous function on the interval $[a, b]$. If F is any antiderivative of f then

$$\int_a^b f(x) \, dx = F(b) - F(a).$$

EXAMPLE 4.5

Calculate

$$\int_0^2 x^2 \, dx.$$

SOLUTION

We use the Fundamental Theorem. In this example, $f(x) = x^2$. We need to find an antiderivative F. From our experience in Section 4.1, we can determine that $F(x) = x^3/3$ will do. Then, by the Fundamental Theorem of Calculus,

$$\int_0^2 x^2 \, dx = F(2) - F(0) = \frac{2^3}{3} - \frac{0^3}{3} = \frac{8}{3}.$$

Notice that this is the same answer that we obtained using Riemann sums in Example 4.4.

EXAMPLE 4.6

Calculate

$$\int_0^\pi \sin x \, dx.$$

SOLUTION

In this example, $f(x) = \sin x$. An antiderivative for f is $F(x) = -\cos x$. Then

$$\int_0^\pi \sin x \, dx = F(\pi) - F(0) = (-\cos \pi) - (-\cos 0) = 1 + 1 = 2.$$

EXAMPLE 4.7

Calculate

$$\int_1^2 e^x - \cos 2x + x^3 - 4x \, dx.$$

SOLUTION

In this example, $f(x) = e^x - \cos 2x + x^3 - 4x$. An antiderivative for f is $F(x) = e^x - (1/2)\sin 2x + x^4/4 - 2x^2$. Therefore

$$\int_1^2 e^x - \cos 2x + x^3 - 4x \, dx = F(2) - F(1)$$

$$= \left(e^2 - \frac{1}{2}\sin(2 \cdot 2) + \frac{2^4}{4} - 2 \cdot 2^2 \right)$$

$$- \left(e^1 - \frac{1}{2}\sin(2 \cdot 1) + \frac{1^4}{4} - 2 \cdot 1^2 \right)$$

$$= (e^2 - e) - \frac{1}{2}[\sin 4 - \sin 2] - \frac{9}{4}.$$

You Try It: Calculate the integral

$$\int_{-3}^{-1} x^3 - \cos x + x \, dx.$$

Math Note: Observe in this last example, in fact in all of our examples, you can use *any* antiderivative of the integrand when you apply the Fundamental Theorem of Calculus. In the last example, we could have taken $F(x) = e^x - (1/2)\sin 2x + x^4/4 - 2x^2 + 5$ and the same answer would have resulted. We invite you to provide the details of this assertion.

Justification for the Fundamental Theorem   Let f be a continuous function on the interval $[a, b]$. Define the *area function* F by

$$F(x) = \text{area under } f, \text{ above the } x\text{-axis, and between 0 and } x.$$

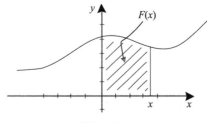

Fig. 4.9

Let us use a pictorial method to calculate the derivative of F. Refer to Fig. 4.9 as you read on. Now

$$\frac{F(x+h) - F(x)}{h} = \frac{[\text{area between } x \text{ and } x+h, \text{ below } f]}{h}$$

$$\approx \frac{f(x) \cdot h}{h}$$

$$= f(x).$$

As $h \to 0$, the approximation in the last display becomes nearer and nearer to equality. So we find that

$$\lim_{h \to 0} \frac{F(x+h) - F(x)}{h} = f(x).$$

But this just says that $F'(x) = f(x)$.

What is the practical significance of this calculation? Suppose that we wish to calculate the area under the curve f, above the x-axis, and between $x = a$ and $x = b$. Obviously this area is $F(b) - F(a)$. See Fig. 4.10. But we also know that

that area is $\int_a^b f(x)\,dx$. We conclude therefore that

$$\int_a^b f(x)\,dx = F(b) - F(a).$$

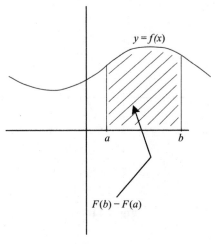

Fig. 4.10

Finally, if G is any other antiderivative for f then $G(x) = F(x) + C$. Hence

$$G(b) - G(a) = [F(b) + C] - [F(a) + C] = F(b) - F(a) = \int_a^b f(x)\,dx.$$

That is the content of the Fundamental Theorem of Calculus.

You Try It: Calculate the area below the curve $y = -x^2 + 2x + 4$ and above the x-axis.

4.3   Signed Area

Without saying so explicitly, we have implicitly assumed in our discussion of area in the last section that our function f is positive, that is its graph lies about the x-axis. But of course many functions do not share that property. We nevertheless would like to be able to calculate areas determined by such functions, and to calculate the corresponding integrals.

   This turns out to be simple to do. Consider the function $y = f(x)$ shown in Fig. 4.11. It is negative on the interval $[a, b]$ and positive on the interval $[b, c]$.

Suppose that we wish to calculate the shaded area as in Fig. 4.12. We can do so by breaking the problem into pieces.

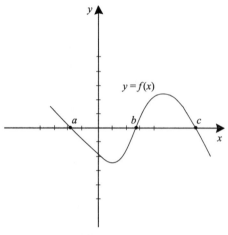

Fig. 4.11

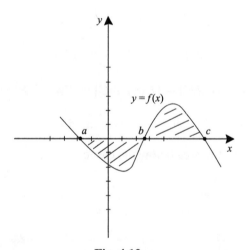

Fig. 4.12

Of course, because $f \geq 0$, the area between $x = b$ and $x = c$ is given by the integral $\int_b^c f(x)\,dx$, just as we have discussed in the last section. But our discussions do not apply directly to the area between $x = a$ and $x = b$. What we can do is instead consider the function $g = -f$. Its graph is shown in Fig. 4.13. Of course g is a positive function on $[a, b]$, except at the endpoints a and b; and the area under g—between $x = a$ and $x = b$—is just the same as the shaded area between

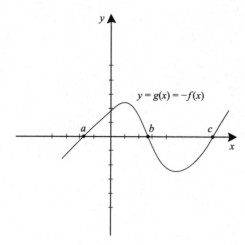

Fig. 4.13

$x = a$ and $x = b$ in Fig. 4.14. That area is

$$\int_a^b g(x)\,dx = -\int_a^b f(x)\,dx.$$

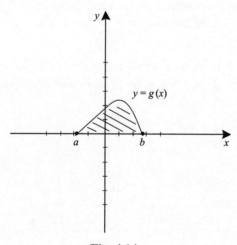

Fig. 4.14

In total, the aggregate shaded area exhibited in Fig. 4.15, over the entire interval $[a, c]$, is

$$-\int_a^b f(x)\,dx + \int_b^c f(x)\,dx.$$

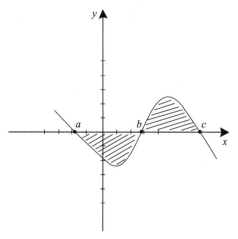

Fig. 4.15

What we have learned is this: If $f(x) < 0$ on the interval under discussion, then the integral of f will be a negative number. If we want to calculate *positive area* then we must interject a minus sign.

Let us nail down our understanding of these ideas by considering some examples.

EXAMPLE 4.8

Calculate the (positive) area, between the graph of $f(x) = x^3 - 2x^2 - 11x + 12$ and the *x*-axis, between $x = -3$ and $x = 4$.

SOLUTION

Consider Fig. 4.16. It was drawn using the technique of Section 3.1, and it plainly shows that f is positive on $[-3, 1]$ and negative on $[1, 4]$. From the

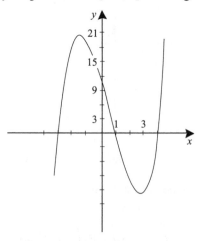

Fig. 4.16

discussion preceding this example, we know then that

$$\text{Area} = \int_{-3}^{1} f(x)\,dx - \int_{1}^{4} f(x)\,dx$$

$$= \int_{-3}^{1} x^3 - 2x^2 - 11x + 12\,dx$$

$$- \int_{1}^{4} x^3 - 2x^2 - 11x + 12\,dx$$

$$= \left(\frac{x^4}{4} - \frac{2x^3}{3} - \frac{11x^2}{2} + 12x \right)\Bigg|_{-3}^{1}$$

$$- \left(\frac{x^4}{4} - \frac{2x^3}{3} - \frac{11x^2}{2} + 12x \right)\Bigg|_{1}^{4}. \qquad (*)$$

Here we are using the standard shorthand

$$F(x)\big|_a^b$$

to stand for

$$F(b) - F(a).$$

Thus we have

$$(*) = \frac{160}{3} + \frac{297}{12}.$$

Notice that, by design, each component of the area has made a positive contribution to the final answer. The total area is then

$$\text{Area} = \frac{937}{12}.$$

EXAMPLE 4.9

Calculate the (positive) area between $f(x) = \sin x$ and the x-axis for $-2\pi \le x \le 2\pi$.

SOLUTION

We observe that $f(x) = \sin x \ge 0$ for $-2\pi \le x \le -\pi$ and $0 \le x \le \pi$. Likewise, $f(x) = \sin x \le 0$ for $-\pi \le x \le 0$ and $\pi \le x \le 2\pi$. As a result

$$\text{Area} = \int_{-2\pi}^{-\pi} \sin x\,dx - \int_{-\pi}^{0} \sin x\,dx$$

$$+ \int_{0}^{\pi} \sin x\,dx - \int_{\pi}^{2\pi} \sin x\,dx.$$

This is easily calculated to equal

$$2 + 2 + 2 + 2 = 8.$$

You Try It: Calculate the (positive) area between $y = x^3 - 6x^2 + 11x - 6$ and the x-axis.

EXAMPLE 4.10

Calculate the signed area between the graph of $y = \cos x + 1/2$ and the x-axis, $-\pi/2 \le x \le \pi$.

SOLUTION

This is easy, because the solution we seek is just the value of the integral:

$$\text{Area} = \int_{-\pi/2}^{\pi} \cos x + \frac{1}{2} \, dx$$

$$= \sin x + \frac{x}{2} \Big|_{-\pi/2}^{\pi}$$

$$= \left[0 + \frac{\pi}{2} \right] - \left[-1 + \frac{-\pi}{4} \right]$$

$$= \frac{3\pi}{4} + 1.$$

Math Note: In the last example, we have counted positive area as positive and negative area as negative. Our calculation shows that the aggregate area is positive. We encourage the reader to draw a graph to make this result plausible.

You Try It: Calculate the actual positive area between the graph of $y = x^2 - 4$, $-5 \le x \le 5$ and the x-axis.

You Try It: Calculate the *signed area* between the graph of $y = x^2 - 4$ and the x-axis, $-4 \le x \le 5$.

4.4 The Area Between Two Curves

Frequently it is useful to find the area *between* two curves. See Fig. 4.17. Following the model that we have set up earlier, we first note that the intersected region has left endpoint at $x = a$ and right endpoint at $x = b$. We partition the interval $[a, b]$ as shown in Fig. 4.18. Call the partition

$$\mathcal{P} = \{x_0, x_1, \ldots, x_k\}.$$

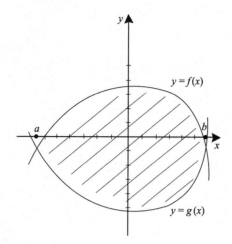

Fig. 4.17

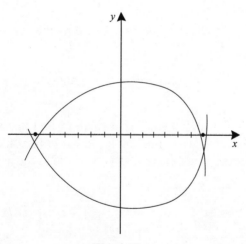

Fig. 4.18

Then, as usual, we erect rectangles over the intervals determined by the partition (Fig. 4.19).

Notice that the upper curve, over the interval $[a, b]$, is $y = f(x)$ and the lower curve is $y = g(x)$ (Fig. 4.17). The sum of the areas of the rectangles is therefore

$$\sum_{j=1}^{k} [f(x_j) - g(x)] \cdot \triangle x.$$

But of course this is a Riemann sum for the integral

$$\int_a^b [f(x) - g(x)] \, dx.$$

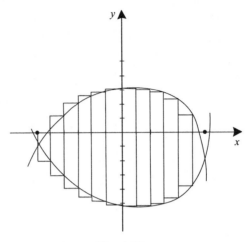

Fig. 4.19

We declare this integral to be the area determined by the two curves.

EXAMPLE 4.11

Find the area between the curves $y = x^2 - 2$ and $y = -(x - 1)^2 + 3$.

SOLUTION

We set the two equations equal and solve to find that the curves intersect at $x = -1$ and $x = 2$. The situation is shown in Fig. 4.20. Notice that $y = -(x - 1)^2 + 3$ is the upper curve and $y = x^2 - 2$ is the lower curve. Thus the desired area is

$$\text{Area} = \int_{-1}^{2} [-(x - 1)^2 + 3] - [x^2 - 2]\, dx$$

$$= \int_{-1}^{2} -2x^2 + 2x + 4\, dx$$

$$= \left. \frac{-2x^3}{3} + x^2 + 4x \right|_{-1}^{2}$$

$$= \left[\frac{-16}{3} + 4 + 8 \right] - \left[\frac{2}{3} + 1 - 4 \right]$$

$$= 9.$$

The area of the region determined by the two parabolas is 9.

EXAMPLE 4.12

Find the area between $y = \sin x$ and $y = \cos x$ for $\pi/4 \le x \le 5\pi/4$.

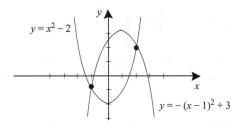

Fig. 4.20

SOLUTION

   On the given interval, $\sin x \geq \cos x$. See Fig. 4.21. Thus the area we wish to compute is

$$\text{Area} = \int_{\pi/4}^{5\pi/4} [\sin x - \cos x]\, dx$$

$$= [-\cos x - \sin x]_{x=\pi/4}^{x=5\pi/4}$$

$$= \left[-\left(-\frac{\sqrt{2}}{2}\right) - \left(-\frac{\sqrt{2}}{2}\right)\right] - \left[-\frac{\sqrt{2}}{2} - \frac{\sqrt{2}}{2}\right]$$

$$= 2\sqrt{2}.$$

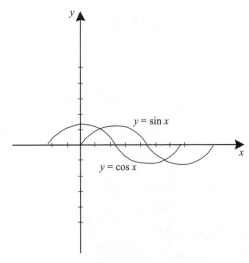

Fig. 4.21

You Try It: Calculate the area between $y = \sin x$ and $y = \cos x$, $-\pi \leq x \leq 2\pi$.

You Try It: Calculate the area between $y = x^2$ and $y = 3x + 4$.

4.5 Rules of Integration

We have been using various rules of integration without enunciating them explicitly. It is well to have them recorded for future reference.

4.5.1 LINEAR PROPERTIES

I. If f, g are continuous functions on $[a, b]$ then

$$\int_a^b f(x) + g(x)\, dx = \int_a^b f(x)\, dx + \int_a^b g(x)\, dx.$$

II. If f is a continuous function on $[a, b]$ and c is a constant then

$$\int_a^b cf(x)\, dx = c \int_a^b f(x)\, dx.$$

4.5.2 ADDITIVITY

III. If f is a continuous on $[a, c]$ and $a < b < c$ then

$$\int_a^b f(x)\, dx + \int_b^c f(x)\, dx = \int_a^c f(x)\, dx.$$

You Try It: Calculate

$$\int_1^3 4x^3 - 3x^2 + 7x - 12 \cos x \, dx.$$

Exercises

1. Calculate each of the following antiderivatives:

(a)   Antiderivative of $x^2 - \sin x$

(b)   Antiderivative of $e^{3x} + x^4 - 2$

(c)   Antiderivative of $3t^2 + (\ln t / t)$

(d)   Antiderivative of $\tan x - \cos x + \sin 3x$

(e)   Antiderivative of $\cos 3x + \sin 4x + 1$

(f)   Antiderivative of $(\cos x) \cdot e^{\sin x}$

2. Calculate each of the following indefinite integrals:

(a) $\int x \sin x^2 \, dx$

(b) $\int (3/x) \ln x^2 \, dx$

(c) $\int \sin x \cdot \cos x \, dx$

(d) $\int \tan x \cdot \ln \cos x \, dx$

(e) $\int \sec^2 x \cdot e^{\tan x} \, dx$

(f) $\int (2x + 1) \cdot (x^2 + x + 7)^{43} \, dx$

3. Use Riemann sums to calculate each of the following integrals:

(a) $\int_1^2 x^2 + x \, dx$

(b) $\int_{-1}^1 (-x^2/3) \, dx$

4. Use the Fundamental Theorem of Calculus to evaluate each of the following integrals:

(a) $\int_1^3 x^2 - 4x^3 + 7 \, dx$

(b) $\int_2^6 x e^{x^2} - \sin x \cos x \, dx$

(c) $\int_1^4 (\ln x/x) + x \sin x^2 \, dx$

(d) $\int_1^2 \tan x - x^2 \cos x^3 \, dx$

(e) $\int_1^e (\ln x^2/x) \, dx$

(f) $\int_4^8 x^2 \cdot \cos x^3 \sin x^3 \, dx$

5. Calculate the area under the given function and above the x-axis over the indicated interval.

(a) $f(x) = x^2 + x + 6$ $[2, 5]$

(b) $g(x) = \sin x \cos x$ $[0, \pi/4]$

(c) $h(x) = x e^{x^2}$ $[1, 2]$

(d) $k(x) = \ln x/x$ $[1, e]$

6. Draw a careful sketch of each function on the given interval, indicating subintervals where area is positive and area is negative.

(a) $f(x) = x^3 + 3x$ $[-2, 2]$

(b) $g(x) = \sin 3x \cos 3x$ $[-2\pi, 2\pi]$

(c) $h(x) = \ln x/x$ $[1/2, e]$

(d) $m(x) = x^3 e^{x^4}$ $[-3, 3]$

7. For each function in Exercise 6, calculate the positive area between the graph of the given function and the x-axis over the indicated interval.

8. In each part of Exercise 6, calculate the *signed area* between the graph of the given function and the x-axis over the indicated interval.

9. Calculate the area between the two given curves over the indicated interval.

   (a) $f(x) = 2x^2 - 4,$ $\quad g(x) = -3x^2 + 10 \quad -1 \le x \le 1$

   (b) $f(x) = x^2,$ $\quad g(x) = x^3 \quad 0 \le x \le 1$

   (c) $f(x) = 2x,$ $\quad g(x) = -x^2 + 3 \quad -3 \le x \le 1$

   (d) $f(x) = \ln x,$ $\quad g(x) = x \quad 1 \le x \le e$

   (e) $f(x) = \sin x,$ $\quad g(x) = x \quad 0 \le x \le \pi/4$

   (f) $f(x) = e^x,$ $\quad g(x) = x \quad 0 \le x \le 3$

10. Calculate the area enclosed by the two given curves.

    (a) $f(x) = x,$ $\quad g(x) = x^2$

    (b) $f(x) = \sqrt{x},$ $\quad g(x) = x^2$

    (c) $f(x) = x^4,$ $\quad g(x) = 3x^2$

    (d) $f(x) = x^4,$ $\quad g(x) = -2x^2 + 3$

    (e) $f(x) = x^4 - 2,$ $\quad g(x) = -x^4 + 2$

    (f) $f(x) = 2x,$ $\quad g(x) = -x^2 + 3$

CHAPTER 5

Indeterminate Forms

5.1   l'Hôpital's Rule

5.1.1   INTRODUCTION

Consider the limit

$$\lim_{x \to c} \frac{f(x)}{g(x)}. \qquad (*)$$

If $\lim_{x \to c} f(x)$ exists and $\lim_{x \to c} g(x)$ exists and is not zero then the limit $(*)$ is straightforward to evaluate. However, as we saw in Theorem 2.3, when $\lim_{x \to c} g(x) = 0$ then the situation is more complicated (especially when $\lim_{x \to c} f(x) = 0$ as well).

For example, if $f(x) = \sin x$ and $g(x) = x$ then the limit of the quotient as $x \to 0$ exists and equals 1. However if $f(x) = x$ and $g(x) = x^2$ then the limit of the quotient as $x \to 0$ does not exist.

In this section we learn a rule for evaluating indeterminate forms of the type $(*)$ when either $\lim_{x \to c} f(x) = \lim_{x \to c} g(x) = 0$ or $\lim_{x \to c} f(x) = \lim_{x \to c} g(x) = \infty$. Such limits, or "forms," are considered indeterminate because the limit of the quotient might actually exist and be finite or might not exist; one cannot analyze such a form by elementary means.

5.1.2 L'HÔPITAL'S RULE

Theorem 5.1 (**l'Hôpital's Rule**)

Let $f(x)$ and $g(x)$ be differentiable functions on $(a, c) \cup (c, b)$. If

$$\lim_{x \to c} f(x) = \lim_{x \to c} g(x) = 0$$

then

$$\lim_{x \to c} \frac{f(x)}{g(x)} = \lim_{x \to c} \frac{f'(x)}{g'(x)},$$

provided this last limit exists as a finite or infinite limit.

Let us learn how to use this new result.

EXAMPLE 5.1

Evaluate

$$\lim_{x \to 1} \frac{\ln x}{x^2 + x - 2}.$$

SOLUTION

We first notice that both the numerator and denominator have limit zero as x tends to 1. Thus the quotient is indeterminate at 1 and of the form $0/0$. l'Hôpital's Rule therefore applies and the limit equals

$$\lim_{x \to 1} \frac{(d/dx)(\ln x)}{(d/dx)(x^2 + x - 2)},$$

provided this last limit exists. The last limit is

$$\lim_{x \to 1} \frac{1/x}{2x + 1} = \lim_{x \to 1} \frac{1}{2x^2 + x}.$$

Therefore we see that

$$\lim_{x \to 1} \frac{\ln x}{x^2 + x - 2} = \frac{1}{3}.$$

You Try It: Apply l'Hôpital's Rule to the limit $\lim_{x \to 2} \sin(\pi x)/(x^2 - 4)$.

You Try It: Use l'Hôpital's Rule to evaluate $\lim_{h \to 0}(\sin h/h)$ and $\lim_{h \to 0}(\cos h - 1/h)$. These limits are important in the theory of calculus.

EXAMPLE 5.2

Evaluate the limit

$$\lim_{x \to 0} \frac{x^3}{x - \sin x}.$$

SOLUTION

As $x \to 0$ both numerator and denominator tend to zero, so the quotient is indeterminate at 0 of the form 0/0. Thus l'Hôpital's Rule applies. Our limit equals

$$\lim_{x \to 0} \frac{(d/dx)x^3}{(d/dx)(x - \sin x)},$$

provided that this last limit exists. It equals

$$\lim_{x \to 0} \frac{3x^2}{1 - \cos x}.$$

This is another indeterminate form. So we must again apply l'Hôpital's Rule. The result is

$$\lim_{x \to 0} \frac{6x}{\sin x}.$$

This is again indeterminate; another application of l'Hôpital's Rule gives us finally

$$\lim_{x \to 0} \frac{6}{\cos x} = 6.$$

We conclude that the original limit equals 6.

You Try It: Apply l'Hôpital's Rule to the limit $\lim_{x \to 0} x / [1/\ln |x|]$.

Indeterminate Forms Involving ∞ We handle indeterminate forms involving infinity as follows: Let $f(x)$ and $g(x)$ be differentiable functions on $(a, c) \cup (c, b)$. If

$$\lim_{x \to c} f(x) \quad \text{and} \quad \lim_{x \to c} g(x)$$

both exist and equal $+\infty$ or $-\infty$ (they may have the same sign or different signs) then

$$\lim_{x \to c} \frac{f(x)}{g(x)} = \lim_{x \to c} \frac{f'(x)}{g'(x)},$$

provided this last limit exists either as a finite or infinite limit.
   Let us look at some examples.

EXAMPLE 5.3

Evaluate the limit

$$\lim_{x \to 0} x^3 \cdot \ln |x|.$$

SOLUTION

This may be rewritten as

$$\lim_{x \to 0} \frac{\ln |x|}{1/x^3}.$$

Notice that the numerator tends to $-\infty$ and the denominator tends to $\pm\infty$ as $x \to 0$. Thus the quotient is indeterminate at 0 of the form $-\infty/+\infty$. So we may apply l'Hôpital's Rule for infinite limits to see that the limit equals

$$\lim_{x \to 0} \frac{1/x}{-3x^{-4}} = \lim_{x \to 0} -x^3/3 = 0.$$

Yet another version of l'Hôpital's Rule, this time for unbounded intervals, is this: Let f and g be differentiable functions on an interval of the form $[A, +\infty)$. If $\lim_{x \to +\infty} f(x) = \lim_{x \to +\infty} g(x) = 0$ or if $\lim_{x \to +\infty} f(x) = \pm\infty$ and $\lim_{x \to +\infty} g(x) = \pm\infty$, then

$$\lim_{x \to +\infty} \frac{f(x)}{g(x)} = \lim_{x \to +\infty} \frac{f'(x)}{g'(x)}$$

provided that this last limit exists either as a finite or infinite limit. The same result holds for f and g defined on an interval of the form $(-\infty, B]$ and for the limit as $x \to -\infty$.

EXAMPLE 5.4

Evaluate

$$\lim_{x \to +\infty} \frac{x^4}{e^x}.$$

SOLUTION

We first notice that both the numerator and the denominator tend to $+\infty$ as $x \to +\infty$. Thus the quotient is indeterminate at $+\infty$ of the form $+\infty/+\infty$. Therefore the new version of l'Hôpital's Rule applies and our limit equals

$$\lim_{x \to +\infty} \frac{4x^3}{e^x}.$$

Again the numerator and denominator tend to $+\infty$ as $x \to +\infty$, so we once more apply l'Hôpital's Rule. The limit equals

$$\lim_{x \to +\infty} \frac{12x^2}{e^x} = 0.$$

We must apply l'Hôpital's Rule two more times. We first obtain

$$\lim_{x \to +\infty} \frac{24x}{e^x}$$

and then

$$\lim_{x \to +\infty} \frac{24}{e^x}.$$

We conclude that

$$\lim_{x \to +\infty} \frac{x^4}{e^x} = 0.$$

You Try It: Evaluate the limit $\lim_{x \to +\infty} \left(\dfrac{e^x}{x \ln x} \right)$.

You Try It: Evaluate the limit $\lim_{x \to -\infty} x^4 \cdot e^x$.

EXAMPLE 5.5

Evaluate the limit

$$\lim_{x \to -\infty} \frac{\sin(2/x)}{\sin(5/x)}.$$

SOLUTION

We note that both numerator and denominator tend to 0, so the quotient is indeterminate at $-\infty$ of the form $0/0$. We may therefore apply l'Hôpital's Rule. Our limit equals

$$\lim_{x \to -\infty} \frac{(-2/x^2)\cos(2/x)}{(-5/x^2)\cos(5/x)}.$$

This in turn simplifies to

$$\lim_{x \to -\infty} \frac{2\cos(2/x)}{5\cos(5/x)} = \frac{2}{5}.$$

l'Hôpital's Rule also applies to one-sided limits. Here is an example.

EXAMPLE 5.6

Evaluate the limit

$$\lim_{x \to 0^+} \frac{\sin \sqrt{x}}{\sqrt{x}}.$$

SOLUTION

Both numerator and denominator tend to zero so the quotient is indeterminate at 0 of the form $0/0$. We may apply l'Hôpital's Rule; differentiating numerator and denominator, we find that the limit equals

$$\lim_{x \to 0^+} \frac{[\cos \sqrt{x}] \cdot (1/2)x^{-1/2}}{(1/2)x^{-1/2}} = \lim_{x \to 0^+} \cos \sqrt{x}$$

$$= 1.$$

You Try It: How can we apply l'Hôpital's Rule to evaluate $\lim_{x \to 0^+} x \cdot \ln x$?

5.2 Other Indeterminate Forms

5.2.1 INTRODUCTION

By using some algebraic manipulations, we can reduce a variety of indeterminate limits to expressions which can be treated by l'Hôpital's Rule. We explore some of these techniques in this section.

5.2.2 WRITING A PRODUCT AS A QUOTIENT

The technique of the first example is a simple one, but it is used frequently.

EXAMPLE 5.7

Evaluate the limit

$$\lim_{x \to -\infty} x^2 \cdot e^{3x}.$$

SOLUTION

Notice that $x^2 \to +\infty$ while $e^{3x} \to 0$. So the limit is indeterminate of the form $0 \cdot \infty$. We rewrite the limit as

$$\lim_{x \to -\infty} \frac{x^2}{e^{-3x}}.$$

Now both numerator and denominator tend to infinity and we may apply l'Hôpital's Rule. The result is that the limit equals

$$\lim_{x \to -\infty} \frac{2x}{-3e^{-3x}}.$$

Again the numerator and denominator both tend to infinity so we apply l'Hôpital's Rule to obtain:

$$\lim_{x \to -\infty} \frac{2}{9e^{-3x}}.$$

It is clear that the limit of this last expression is zero. We conclude that

$$\lim_{x \to -\infty} x \cdot e^{3x} = 0.$$

You Try It: Evaluate the limit $\lim_{x \to +\infty} e^{-\sqrt{x}} \cdot x$.

5.2.3 THE USE OF THE LOGARITHM

The natural logarithm can be used to reduce an expression involving exponentials to one involving a product or a quotient.

EXAMPLE 5.8

Evaluate the limit

$$\lim_{x \to 0^+} x^x.$$

SOLUTION

We study the limit of $f(x) = x^x$ by considering $\ln f(x) = x \cdot \ln x$. We rewrite this as

$$\lim_{x \to 0^+} \ln f(x) = \lim_{x \to 0^+} \frac{\ln x}{1/x}.$$

Both numerator and denominator tend to $\pm\infty$, so the quotient is indeterminate of the form $-\infty/\infty$. Thus l'Hôpital's Rule applies. The limit equals

$$\lim_{x \to 0^+} \frac{1/x}{-1/x^2} = \lim_{x \to 0^+} -x = 0.$$

Now the only way that $\ln f(x)$ can tend to zero is if $f(x) = x^x$ tends to 1. We conclude that

$$\lim_{x \to 0^+} x^x = 1.$$

EXAMPLE 5.9

Evaluate the limit

$$\lim_{x \to 0} (1 + x^2)^{\ln |x|}.$$

SOLUTION

Let $f(x) = (1 + x^2)^{\ln |x|}$ and consider $\ln f(x) = \ln |x| \cdot \ln(1 + x^2)$. This expression is indeterminate of the form $-\infty \cdot 0$.

We rewrite it as

$$\lim_{x \to 0} \frac{\ln(1 + x^2)}{1/\ln |x|},$$

so that both the numerator and denominator tend to 0. So l'Hôpital's Rule applies and we have

$$\lim_{x \to 0} \ln f(x) = \lim_{x \to 0} \frac{2x/(1 + x^2)}{-1/[x \ln^2(|x|)]} = \lim_{x \to 0} -\frac{2x^2 \ln^2(|x|)}{(1 + x^2)}.$$

The numerator tends to 0 (see Example 5.3) and the denominator tends to 1. Thus

$$\lim_{x \to 0} \ln f(x) = 0.$$

But the only way that ln $f(x)$ can tend to zero is if $f(x)$ tends to 1. We conclude that

$$\lim_{x \to 0} (1 + x^2)^{\ln |x|} = 1.$$

You Try It: Evaluate the limit $\lim_{x \to 0^+} (1/x)^x$.

You Try It: Evaluate the limit $\lim_{x \to 0^+} (1 + x)^{1/x}$. In fact this limit gives an important way to define Euler's constant e (see Sections 1.9 and 6.2.3).

5.2.4   PUTTING TERMS OVER A COMMON DENOMINATOR

Many times a simple algebraic manipulation involving fractions will put a limit into a form which can be studied using l'Hôpital's Rule.

EXAMPLE 5.10

Evaluate the limit

$$\lim_{x \to 0} \left[\frac{1}{\sin 3x} - \frac{1}{3x} \right].$$

SOLUTION

We put the fractions over a common denominator to rewrite our limit as

$$\lim_{x \to 0} \left[\frac{3x - \sin 3x}{3x \cdot \sin 3x} \right].$$

Both numerator and denominator vanish as $x \to 0$. Thus the quotient has indeterminate form $0/0$. By l'Hôpital's Rule, the limit is therefore equal to

$$\lim_{x \to 0} \frac{3 - 3\cos 3x}{3 \sin 3x + 9x \cos 3x}.$$

This quotient is still indeterminate; we apply l'Hôpital's Rule again to obtain

$$\lim_{x \to 0} \frac{9 \sin 3x}{18 \cos 3x - 27x \sin 3x} = 0.$$

EXAMPLE 5.11

Evaluate the limit

$$\lim_{x \to 0} \left[\frac{1}{4x} - \frac{1}{e^{4x} - 1} \right].$$

SOLUTION

The expression is indeterminate of the form $\infty - \infty$. We put the two fractions over a common denominator to obtain

$$\lim_{x \to 0} \frac{e^{4x} - 1 - 4x}{4x(e^{4x} - 1)}.$$

Notice that the numerator and denominator both tend to zero as $x \to 0$, so this is indeterminate of the form $0/0$. Therefore l'Hôpital's Rule applies and our limit equals

$$\lim_{x \to 0} \frac{4e^{4x} - 4}{4e^{4x}(1 + 4x) - 4}.$$

Again the numerator and denominator tend to zero and we apply l'Hôpital's Rule; the limit equals

$$\lim_{x \to 0} \frac{16e^{4x}}{16e^{4x}(2 + 4x)} = \frac{1}{2}.$$

You Try It: Evaluate the limit $\lim_{x \to 0} \left(\dfrac{1}{\cos x - 1} \right) + \left(\dfrac{2}{x^2} \right)$.

5.2.5   OTHER ALGEBRAIC MANIPULATIONS

Sometimes a factorization helps to clarify a subtle limit:

EXAMPLE 5.12

Evaluate the limit

$$\lim_{x \to +\infty} \left[x^2 - (x^4 + 4x^2 + 5)^{1/2} \right].$$

SOLUTION

The limit as written is of the form $\infty - \infty$. We rewrite it as

$$\lim_{x \to +\infty} x^2 \left[1 - (1 + 4x^{-2} + 5x^{-4})^{1/2} \right] = \lim_{x \to +\infty} \frac{1 - (1 + 4x^{-2} + 5x^{-4})^{1/2}}{x^{-2}}.$$

Notice that both the numerator and denominator tend to zero, so it is now indeterminate of the form $0/0$. We may thus apply l'Hôpital's Rule. The result is that the limit equals

$$\lim_{x \to +\infty} \frac{(-1/2)(1 + 4x^{-2} + 5x^{-4})^{-1/2} \cdot (-8x^{-3} - 20x^{-5})}{-2x^{-3}}$$

$$= \lim_{x \to +\infty} -(1 + 4x^{-2} + 5x^{-4})^{-1/2} \cdot (2 + 5x^{-2}).$$

Since this last limit is -2, we conclude that

$$\lim_{x \to +\infty} \left[x^2 - (x^4 + 4x^2 + 5)^{1/2} \right] = -2.$$

EXAMPLE 5.13

Evaluate

$$\lim_{x \to -\infty} \left[e^{-x} - (e^{-3x} - x^4)^{1/3} \right].$$

SOLUTION

First rewrite the limit as

$$\lim_{x \to -\infty} e^{-x} \left[1 - (1 - x^4 e^{3x})^{1/3} \right] = \lim_{x \to -\infty} \frac{1 - (1 - x^4 e^{3x})^{1/3}}{e^x}.$$

Notice that both the numerator and denominator tend to zero (here we use the result analogous to Example 5.7 that $x^4 e^{3x} \to 0$). So our new expression is indeterminate of the form $0/0$. l'Hôpital's Rule applies and our limit equals

$$\lim_{x \to -\infty} \frac{-(1/3)(1 - x^4 e^{3x})^{-2/3} \cdot (-4x^3 \cdot e^{3x} - x^4 \cdot 3e^{3x})}{e^x}$$

$$= \lim_{x \to -\infty} (1/3)(1 - x^4 e^{3x})^{-2/3}(4x^3 \cdot e^{2x} + 3x^4 \cdot e^{2x}).$$

Just as in Example 5.7, $x^4 \cdot e^{3x}$, $x^3 \cdot e^{2x}$, and $x^4 \cdot e^{2x}$ all tend to zero. We conclude that our limit equals 0.

You Try It: Evaluate $\lim_{x \to +\infty} [\sqrt{x + 1} - \sqrt{x}]$.

5.3 Improper Integrals: A First Look

5.3.1 INTRODUCTION

The theory of the integral that we learned earlier enables us to integrate a continuous function $f(x)$ on a closed, bounded interval $[a, b]$. See Fig. 5.1. However, it is frequently convenient to be able to integrate an unbounded function, or a function defined on an unbounded interval. In this section and the next we learn to do so, and we see some applications of this new technique. The basic idea is that the integral of an unbounded function is the limit of integrals of bounded functions; likewise, the integral of a function on an unbounded interval is the limit of the integral on bounded intervals.

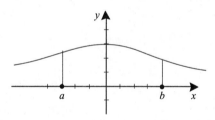

Fig. 5.1

5.3.2   INTEGRALS WITH INFINITE INTEGRANDS

Let f be a continuous function on the interval $[a, b)$ which is unbounded as $x \to b^-$ (see Fig. 5.2). The integral

$$\int_a^b f(x)\, dx$$

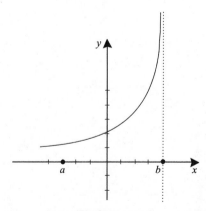

Fig. 5.2

is then called an *improper integral with infinite integrand* at b. We often just say "improper integral" because the *source* of the improperness will usually be clear from context. The next definition tells us how such an integral is evaluated.

If

$$\int_a^b f(x)\, dx$$

is an improper integral with infinite integrand at b then the value of the integral is defined to be

$$\lim_{\epsilon \to 0^+} \int_a^{b-\epsilon} f(x)\, dx,$$

provided that this limit exists. See Fig. 5.3.

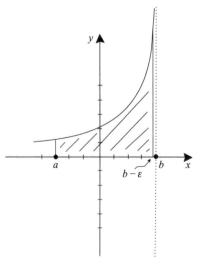

Fig. 5.3

EXAMPLE 5.14

Evaluate the integral

$$\int_2^8 4(8-x)^{-1/3}\,dx.$$

SOLUTION

The integral

$$\int_2^8 4(8-x)^{-1/3}\,dx$$

is an improper integral with infinite integrand at 8. According to the definition, the value of this integral is

$$\lim_{\epsilon \to 0^+}\int_2^{8-\epsilon} 4(8-x)^{-1/3}\,dx,$$

provided the limit exists. Since the integrand is continuous on the interval $[2, 8-\epsilon]$, we may calculate this last integral directly. We have

$$\lim_{\epsilon \to 0^+}\left[-6(8-x)^{2/3}\right]\Big|_2^{8-\epsilon} = \lim_{\epsilon \to 0^+} -6\left[\epsilon^{2/3} - 6^{2/3}\right].$$

This limit is easy to evaluate: it equals $6^{5/3}$. We conclude that the integral is convergent and

$$\int_2^8 4(8-x)^{-1/3}\,dx = 6^{5/3}.$$

EXAMPLE 5.15

Analyze the integral

$$\int_2^3 (x-3)^{-2}\, dx.$$

SOLUTION

This is an improper integral with infinite integrand at 3. We evaluate this integral by considering

$$\lim_{\epsilon \to 0^+} \int_2^{3-\epsilon} (x-3)^{-2}\, dx = \lim_{\epsilon \to 0^+} -(x-3)^{-1}\Big|_2^{3-\epsilon}$$

$$= \lim_{\epsilon \to 0^+} \left[\epsilon^{-1} - 1^{-1} \right].$$

This last limit is $+\infty$. We therefore conclude that the improper integral *diverges*.

You Try It: Evaluate the improper integral $\int_{-2}^{-1}(dx/(x+1)^{4/5})\, dx$.

Improper integrals with integrand which is infinite at the left endpoint of integration are handled in a manner similar to the right endpoint case:

EXAMPLE 5.16

Evaluate the integral

$$\int_0^{1/2} \frac{1}{x \cdot \ln^2 x}\, dx.$$

SOLUTION

This integral is improper with infinite integrand at 0. The value of the integral is defined to be

$$\lim_{\epsilon \to 0^+} \int_\epsilon^{1/2} \frac{1}{x \cdot \ln^2 x}\, dx,$$

provided that this limit exists.

Since $1/(x \ln^2 x)$ is continuous on the interval $[\epsilon, 1/2]$ for $\epsilon > 0$, this last integral can be evaluated directly and will have a finite real value. For clarity, write $\varphi(x) = \ln x$, $\varphi'(x) = 1/x$. Then the (indefinite) integral becomes

$$\int \frac{\varphi'(x)}{\varphi^2(x)}\, dx.$$

Clearly the antiderivative is $-1/\varphi(x)$. Thus we see that

$$\lim_{\epsilon \to 0^+} \int_\epsilon^{1/2} \frac{1}{x \cdot \ln^2 x}\, dx = \lim_{\epsilon \to 0^+} -\frac{1}{\ln x}\Big|_\epsilon^{1/2} = \lim_{\epsilon \to 0^+} \left(\left[-\frac{1}{\ln(1/2)} \right] - \left[-\frac{1}{\ln \epsilon} \right] \right).$$

Now as $\epsilon \to 0^+$ we have $\ln \epsilon \to -\infty$ hence $1/\ln \epsilon \to 0$. We conclude that the improper integral *converges* to $1/\ln 2$.

You Try It: Evaluate the improper integral $\int_{-2}^{0} 1/(x+2)^{-1/2}\, dx$.

Many times the integrand has a singularity in the middle of the interval of integration. In these circumstances we divide the integral into two pieces for each of which the integrand is infinite at one endpoint, and evaluate each piece separately.

EXAMPLE 5.17

Evaluate the improper integral

$$\int_{-4}^{4} 4(x+1)^{-1/5}\, dx.$$

SOLUTION

The integrand is unbounded as x tends to -1. Therefore we evaluate separately the two improper integrals

$$\int_{-4}^{-1} 4(x+1)^{-1/5}\, dx \ \text{ and } \ \int_{-1}^{4} 4(x+1)^{-1/5}\, dx.$$

The first of these has the value

$$\lim_{\epsilon \to 0^+} \int_{-4}^{-1-\epsilon} 4(x+1)^{-1/5}\, dx = \lim_{\epsilon \to 0^+} 5(x+1)^{4/5}\Big|_{-4}^{-1-\epsilon}$$

$$= \lim_{\epsilon \to 0^+} 5\big\{(-\epsilon)^{4/5} - (-3)^{4/5}\big\}$$

$$= -5 \cdot 3^{4/5}$$

The second integral has the value

$$\lim_{\epsilon \to 0^+} \int_{-1+\epsilon}^{4} 4(x+1)^{-1/5}\, dx = \lim_{\epsilon \to 0^+} 5(x+1)^{4/5}\Big|_{-1+\epsilon}^{4}$$

$$= \lim_{\epsilon \to 0^+} 5\big\{5^{4/5} - \epsilon^{4/5}\big\}$$

$$= 5^{9/5}.$$

We conclude that the original integral converges and

$$\int_{-4}^{4} 4(x+1)^{-1/5}\, dx = \int_{-4}^{-1} 4(x+1)^{-1/5}\, dx + \int_{-1}^{4} 4(x+1)^{-1/5}\, dx$$

$$= -5 \cdot 3^{4/5} + 5^{9/5}.$$

You Try It: Evaluate the improper integral $\int_{-4}^{3} x^{-1}\, dx$.

It is dangerous to try to save work by not dividing the integral at the singularity. The next example illustrates what can go wrong.

EXAMPLE 5.18

Evaluate the improper integral

$$\int_{-2}^{2} x^{-4}\, dx.$$

SOLUTION

What we *should* do is divide this problem into the two integrals

$$\int_{-2}^{0} x^{-4}\, dx \quad \text{and} \quad \int_{0}^{2} x^{-4}\, dx. \qquad (*)$$

Suppose that instead we try to save work and just antidifferentiate:

$$\int_{-2}^{2} x^{-4}\, dx = -\frac{1}{3}x^{-3}\Big|_{-2}^{2} = -\frac{1}{12}.$$

A glance at Fig. 5.4 shows that something is wrong. The function x^{-4} is positive hence its integral should be positive too. However, since we used an incorrect method, we got a negative answer.

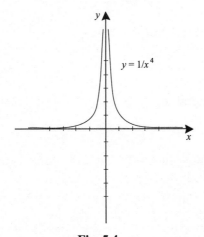

$y = 1/x^4$

Fig. 5.4

In fact each of the integrals in line $(*)$ diverges, so *by definition* the improper integral

$$\int_{-2}^{2} x^{-4}\, dx$$

diverges.

EXAMPLE 5.19

Analyze the integral

$$\int_0^1 \frac{1}{x(1-x)^{1/2}}\,dx.$$

SOLUTION

The key idea is that we can only handle one singularity at a time. This integrand is singular at both endpoints 0 and 1. Therefore we divide the domain of integration somewhere in the middle—at $1/2$ say (it does not really matter where we divide)—and treat the two singularities separately.

First we treat the integral

$$\int_0^{1/2} \frac{1}{x(1-x)^{1/2}}\,dx.$$

Since the integrand has a singularity at 0, we consider

$$\lim_{\epsilon \to 0^+} \int_\epsilon^{1/2} \frac{1}{x(1-x)^{1/2}}\,dx.$$

This is a tricky integral to evaluate directly. But notice that

$$\frac{1}{x(1-x)^{1/2}} \geq \frac{1}{x \cdot (1)^{1/2}}$$

when $0 < \epsilon \leq x \leq 1/2$. Thus

$$\int_\epsilon^{1/2} \frac{1}{x(1-x)^{1/2}}\,dx \geq \int_\epsilon^{1/2} \frac{1}{x \cdot (1)^{1/2}}\,dx = \int_\epsilon^{1/2} \frac{1}{x}\,dx.$$

We evaluate the integral: it equals $\ln(1/2) - \ln \epsilon$. Finally,

$$\lim_{\epsilon \to 0^+} -\ln \epsilon = +\infty.$$

The first of our integrals therefore diverges.

But the full integral

$$\int_0^1 \frac{1}{x(1-x)^{1/2}}\,dx$$

converges if and only if each of the component integrals

$$\int_0^{1/2} \frac{1}{x(1-x)^{1/2}}\,dx$$

and

$$\int_{1/2}^1 \frac{1}{x(1-x)^{1/2}}\,dx$$

converges. Since the first integral diverges, we conclude that the original integral diverges as well.

You Try It: Calculate $\int_{-2}^{3}(2x)^{-1/3}\,dx$ as an improper integral.

5.3.3   AN APPLICATION TO AREA

Suppose that f is a non-negative, continuous function on the interval $(a, b]$ which is unbounded as $x \to a^+$. Look at Fig. 5.5. Let us consider the area under the graph of f and above the x-axis over the interval $(a, b]$. The area of the part of the region over the interval $[a + \epsilon, b]$, $\epsilon > 0$, is

$$\int_{a+\epsilon}^{b} f(x)\,dx.$$

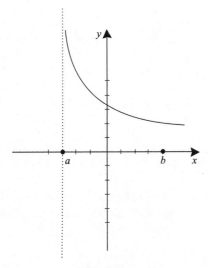

Fig. 5.5

Therefore it is natural to consider the area of the entire region, over the interval $(a, b]$, to be

$$\lim_{\epsilon \to 0^+} \int_{a+\epsilon}^{b} f(x)\,dx.$$

This is just the improper integral

$$\text{Area} = \int_{a}^{b} f(x)\,dx.$$

EXAMPLE 5.20

Calculate the area above the *x*-axis and under the curve

$$y = \frac{1}{x \cdot \ln^{4/3} x}, \qquad 0 < x \le 1/2.$$

SOLUTION

According to the preceding discussion, this area is equal to the value of the improper integral

$$\int_0^{1/2} \frac{1}{x \cdot \ln^{4/3} x} \, dx = \lim_{\epsilon \to 0^+} \int_\epsilon^{1/2} \frac{1}{x \cdot \ln^{4/3} x} \, dx.$$

For clarity we let $\varphi(x) = \ln x$, $\varphi'(x) = 1/x$. Then the (indefinite) integral becomes

$$\int \frac{\varphi'(x)}{\varphi^{4/3}(x)} \, dx = -\frac{3}{\varphi^{1/3}(x)}.$$

Thus

$$\lim_{\epsilon \to 0^+} \int_\epsilon^{1/2} \frac{1}{x \cdot \ln^{4/3} x} \, dx = \lim_{\epsilon \to 0^+} -\frac{3}{\ln^{1/3} x} \bigg|_\epsilon^{1/2}$$

$$= \lim_{\epsilon \to 0^+} \left[\frac{-3}{[-\ln 2]^{1/3}} - \frac{-3}{[\ln \epsilon]^{1/3}} \right].$$

Now as $\epsilon \to 0$ then $\ln \epsilon \to -\infty$ hence $1/[\ln \epsilon]^{1/3} \to 0$. We conclude that our improper integral converges and the area under the curve and above the *x*-axis equals $3/[\ln 2]^{1/3}$.

5.4   More on Improper Integrals

5.4.1   INTRODUCTION

Suppose that we want to calculate the integral of a continuous function $f(x)$ over an unbounded interval of the form $[A, +\infty)$ or $(-\infty, B]$. The theory of the integral that we learned earlier does not cover this situation, and some new concepts are needed. We treat improper integrals on infinite intervals in this section, and give some applications at the end.

5.4.2   THE INTEGRAL ON AN INFINITE INTERVAL

Let f be a continuous function whose domain contains an interval of the form $[A, +\infty)$. The value of the improper integral

$$\int_A^{+\infty} f(x)\, dx$$

is defined to be

$$\lim_{N \to +\infty} \int_A^N f(x)\, dx.$$

Similarly, we have: Let g be a continuous function whose domain contains an interval of the form $(-\infty, B]$. The value of the improper integral

$$\int_{-\infty}^B g(x)\, dx$$

is defined to be

$$\lim_{M \to -\infty} \int_M^B f(x)\, dx.$$

EXAMPLE 5.21

Calculate the improper integral

$$\int_1^{+\infty} x^{-3}\, dx.$$

SOLUTION

We do this problem by evaluating the limit

$$\lim_{N \to +\infty} \int_1^N x^{-3}\, dx = \lim_{N \to +\infty} \left[-(1/2)x^{-2} \Big|_1^N \right]$$

$$= \lim_{N \to +\infty} -(1/2)\left[N^{-2} - 1^{-2} \right]$$

$$= \frac{1}{2}.$$

We conclude that the integral converges and has value $1/2$.

EXAMPLE 5.22

Evaluate the improper integral

$$\int_{-\infty}^{-32} x^{-1/5}\, dx.$$

SOLUTION

We do this problem by evaluating the limit

$$\lim_{M \to -\infty} \int_M^{-32} x^{-1/5} \, dx = \lim_{M \to -\infty} \frac{5}{4} x^{4/5} \Big|_M^{-32}$$

$$= \lim_{M \to -\infty} \frac{5}{4} \big[(-32)^{4/5} - M^{4/5} \big]$$

$$= \lim_{M \to -\infty} \frac{5}{4} \big[16 - M^{4/5} \big].$$

This limit equals $-\infty$. Therefore the integral diverges.

You Try It: Evaluate $\int_1^\infty (1 + x)^{-3} \, dx$.

Sometimes we have occasion to evaluate a doubly infinite integral. We do so by breaking the integral up into two separate improper integrals, each of which can be evaluated with just one limit.

EXAMPLE 5.23

Evaluate the improper integral

$$\int_{-\infty}^\infty \frac{1}{1 + x^2} \, dx.$$

SOLUTION

The interval of integration is $(-\infty, +\infty)$. To evaluate this integral, we break the interval up into two pieces:

$$(-\infty, +\infty) = (-\infty, 0] \cup [0, +\infty).$$

(The choice of zero as a place to break the interval is not important; any other point would do in this example.) Thus we will evaluate separately the integrals

$$\int_0^{+\infty} \frac{1}{1 + x^2} \, dx \quad \text{and} \quad \int_{-\infty}^0 \frac{1}{1 + x^2} \, dx.$$

For the first one we consider the limit

$$\lim_{N \to +\infty} \int_0^N \frac{1}{1 + x^2} \, dx = \lim_{N \to +\infty} \mathrm{Tan}^{-1} x \Big|_0^N$$

$$= \lim_{N \to +\infty} \big[\mathrm{Tan}^{-1} N - \mathrm{Tan}^{-1} 0 \big]$$

$$= \frac{\pi}{2}.$$

The second integral is evaluated similarly:

$$\lim_{M \to -\infty} \int_M^0 \frac{1}{1+x^2}\,dx = \lim_{M \to -\infty} \mathrm{Tan}^{-1}x \Big|_M^0$$

$$= \lim_{M \to -\infty} \left[\mathrm{Tan}^{-1}0 - \mathrm{Tan}^{-1}M\right]$$

$$= \frac{\pi}{2}.$$

Since each of the integrals on the half line is convergent, we conclude that the original improper integral over the entire real line is convergent and that its value is

$$\frac{\pi}{2} + \frac{\pi}{2} = \pi.$$

You Try It: Discuss $\int_1^\infty (1+x)^{-1}\,dx$.

5.4.3   SOME APPLICATIONS

Now we may use improper integrals over infinite intervals to calculate area.

EXAMPLE 5.24
Calculate the area under the curve $y = 1/[x \cdot (\ln x)^4]$ and above the x-axis, $2 \le x < \infty$.

SOLUTION
The area is given by the improper integral

$$\int_2^{+\infty} \frac{1}{x \cdot (\ln x)^4}\,dx = \lim_{N \to +\infty} \int_2^N \frac{1}{x \cdot (\ln x)^4}\,dx.$$

For clarity, we let $\varphi(x) = \ln x$, $\varphi'(x) = 1/x$. Thus the (indefinite) integral becomes

$$\int \frac{\varphi'(x)}{\varphi^4(x)}\,dx = -\frac{1/3}{\varphi^3(x)}.$$

Thus

$$\lim_{N \to +\infty} \int_2^N \frac{1}{x \cdot (\ln x)^4}\,dx = \lim_{N \to +\infty} \left[-\frac{1/3}{\ln^3 x}\right]_2^N$$

$$= \lim_{N \to +\infty} -\left[\frac{1/3}{\ln^3 N} - \frac{1/3}{\ln^3 2}\right]$$

$$= \frac{1/3}{\ln^3 2}.$$

Thus the area under the curve and above the x-axis is $1/(3 \ln^3 2)$.

EXAMPLE 5.25

Because of inflation, the value of a dollar decreases as time goes on. Indeed, this decrease in the value of money is directly related to the continuous compounding of interest. For if one dollar today is invested at 6% continuously compounded interest for ten years then that dollar will have grown to $e^{0.06 \cdot 10}$ = $1.82 (see Section 6.5 for more detail on this matter). This means that a dollar in the currency of ten years from now corresponds to only $e^{-0.06 \cdot 10}$ = $0.55 in today's currency.

Now suppose that a trust is established in your name which pays $2t + 50$ dollars per year for every year in perpetuity, where t is time measured in years (here the present corresponds to time $t = 0$). Assume a constant interest rate of 6%, and that all interest is re-invested. What is the total value, in today's dollars, of all the money that will ever be earned by your trust account?

SOLUTION

Over a short time increment $[t_{j-1}, t_j]$, the value *in today's currency* of the money earned is about

$$(2t_j + 50) \cdot \left(e^{-0.06 \cdot t_j}\right) \cdot \Delta t_j.$$

The corresponding sum over time increments is

$$\sum_j (2t_j + 50) \cdot e^{-0.06 \cdot t_j} \Delta t_j.$$

This in turn is a Riemann sum for the integral

$$\int (2t + 50)e^{-0.06t} \, dt.$$

If we want to calculate the value in today's dollars of all the money earned from now on, in perpetuity, this would be the value of the improper integral

$$\int_0^{+\infty} (2t + 50)e^{-0.06t} \, dt.$$

This value is easily calculated to be $1388.89, rounded to the nearest cent.

You Try It: A trust is established in your name which pays $t + 10$ dollars per year for every year in perpetuity, where t is time measured in years (here the present corresponds to time $t = 0$). Assume a constant interest rate of 4%. What is the total value, in today's dollars, of all the money that will ever be earned by your trust account?

Exercises

1. If possible, use l'Hôpital's Rule to evaluate each of the following limits. In each case, check carefully that the hypotheses of l'Hôpital's Rule apply.

   (a) $\displaystyle\lim_{x\to 0}\frac{\cos x - 1}{x^2 - x^3}$

   (b) $\displaystyle\lim_{x\to 0}\frac{e^{2x} - 1 - 2x}{x^2 + x^4}$

   (c) $\displaystyle\lim_{x\to 0}\frac{\cos x}{x^2}$

   (d) $\displaystyle\lim_{x\to 1}\frac{[\ln x]^2}{(x - 1)}$

   (e) $\displaystyle\lim_{x\to 2}\frac{(x - 2)^3}{\sin(x - 2) - (x - 2)}$

   (f) $\displaystyle\lim_{x\to 1}\frac{e^x - 1}{x - 1}$

2. If possible, use l'Hôpital's Rule to evaluate each of the following limits. In each case, check carefully that the hypotheses of l'Hôpital's Rule apply.

   (a) $\displaystyle\lim_{x\to +\infty}\frac{x^3}{e^x - x^2}$

   (b) $\displaystyle\lim_{x\to +\infty}\frac{\ln x}{x}$

   (c) $\displaystyle\lim_{x\to +\infty}\frac{e^{-x}}{\ln[x/(x + 1)]}$

   (d) $\displaystyle\lim_{x\to +\infty}\frac{\sin x}{e^{-x}}$

   (e) $\displaystyle\lim_{x\to -\infty}\frac{e^x}{1/x}$

   (f) $\displaystyle\lim_{x\to -\infty}\frac{\ln |x|}{e^{-x}}$

3. If possible, use some algebraic manipulations, plus l'Hôpital's Rule, to evaluate each of the following limits. In each case, check carefully that the hypotheses of l'Hôpital's Rule apply.

   (a) $\displaystyle\lim_{x\to +\infty} x^3 e^{-x}$

   (b) $\displaystyle\lim_{x\to +\infty} x \cdot \sin[1/x]$

(c) $\displaystyle\lim_{x\to+\infty} \ln[x/(x+1)] \cdot (x+1)$

(d) $\displaystyle\lim_{x\to+\infty} \ln x \cdot e^{-x}$

(e) $\displaystyle\lim_{x\to-\infty} e^{2x} \cdot x^2$

(f) $\displaystyle\lim_{x\to 0} x \cdot e^{1/x}$

4. Evaluate each of the following improper integrals. In each case, be sure to write the integral as an appropriate limit.

(a) $\displaystyle\int_0^1 x^{-3/4}\,dx$

(b) $\displaystyle\int_1^3 (x-3)^{-4/3}\,dx$

(c) $\displaystyle\int_{-2}^2 \frac{1}{(x+1)^{1/3}}\,dx$

(d) $\displaystyle\int_{-4}^6 \frac{x}{(x-1)(x+2)}\,dx$

(e) $\displaystyle\int_4^8 \frac{x+4}{(x-8)^{1/3}}\,dx$

(f) $\displaystyle\int_0^3 \frac{\sin x}{x^2}\,dx$

5. Evaluate each of the following improper integrals. In each case, be sure to write the integral as an appropriate limit.

(a) $\displaystyle\int_1^\infty e^{-3x}\,dx$

(b) $\displaystyle\int_2^\infty x^2 e^{-x}\,dx$

(c) $\displaystyle\int_0^\infty x \ln x\,dx$

(d) $\displaystyle\int_1^\infty \frac{dx}{1+x^2}$

(e) $\displaystyle\int_1^\infty \frac{dx}{x}$

(f) $\displaystyle\int_{-\infty}^{-1} \frac{dx}{x^2+x}$

CHAPTER 6

Transcendental Functions

6.0 Introductory Remarks

There are two types of functions: polynomial and transcendental. A *polynomial of degree k* is a function of the form $p(x) = a_0 + a_1 x + a_2 x^2 + \cdots + a_k x^k$. Such a polynomial has precisely k roots, and there are algorithms that enable us to solve for those roots. For most purposes, polynomials are the most accessible and easy-to-understand functions. But there are other functions that are important in mathematics and physics. These are the transcendental functions. Among this more sophisticated type of functions are sine, cosine, the other trigonometric functions, and also the logarithm and the exponential. The present chapter is devoted to the study of transcendental functions.

6.1 Logarithm Basics

A convenient, intuitive way to think about the logarithm function is as the inverse to the exponentiation function. Proceeding intuitively, let us consider the function

$$f(x) = 3^x.$$

To operate with this f, we choose an x and take 3 to the power x. For example,

$$f(4) = 3^4 = 3 \cdot 3 \cdot 3 \cdot 3 = 81$$

$$f(-2) = 3^{-2} = \frac{1}{9}$$
$$f(0) = 3^0 = 1.$$

The *inverse* of the function f is the function g which assigns to x the power to which you need to raise 3 to obtain x. For instance,

$$g(9) = 2 \quad \text{because } f(2) = 9$$
$$g(1/27) = -3 \quad \text{because } f(-3) = 1/27$$
$$g(1) = 0 \quad \text{because } f(0) = 1.$$

We usually call the function g the "logarithm to the base 3" and we write $g(x) = \log_3 x$. Logarithms to other bases are defined similarly.

While this approach to logarithms has heuristic appeal, it has many drawbacks: we do not really know what 3^x means when x is not a rational number; we have no way to determine the derivative of f or of g; we have no way to determine the integral of f or of g. Because of these difficulties, we are going to use an entirely new method for studying logarithms. It turns out to be equivalent to the intuitive method described above, and leads rapidly to the calculus results that we need.

6.1.1 A NEW APPROACH TO LOGARITHMS

When you studied logarithms in the past you learned the formula

$$\log(x \cdot y) = \log x + \log y;$$

this says that logs convert multiplication to addition. It turns out that this property alone uniquely determines the logarithm function.

Let $\ell(x)$ be a differentiable function with domain the positive real numbers and whose derivative function $\ell'(x)$ is continuous. Assume that ℓ satisfies the multiplicative law

$$\ell(x \cdot y) = \ell(x) + \ell(y) \tag{$*$}$$

for all positive x and y. Then it must be that $\ell(1) = 0$ and there is a constant C such that

$$\ell'(x) = \frac{C}{x}.$$

In other words

$$\ell(x) = \int_1^x \frac{C}{t}\, dt.$$

A function $\ell(x)$ that satisfies these properties is called a *logarithm function*. The particular logarithm function which satisfies $\ell'(1) = 1$ is called the *natural logarithm*: In other words,

$$\text{natural logarithm} = \ln x = \int_1^x \frac{1}{t}\, dt.$$

For $0 < x < 1$ the value of $\ln x$ is the *negative* of the actual area between the graph and the x-axis. This is so because the limits of integration, x and 1, occur in reverse order: $\ln x = \int_1^x (1/t)\, dt$ with $x < 1$.

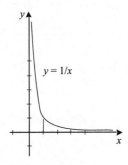

Fig. 6.1

Notice the following simple properties of $\ln x$ which can be determined from looking at Fig. 6.1:

(i)   When $x > 1$, $\ln x > 0$ (after all, $\ln x$ is an *area*).

(ii)   When $x = 1$, $\ln x = 0$.

(iii)   When $0 < x < 1$, $\ln x < 0$

$$\left(\text{since} \int_1^x \frac{1}{t}\, dt = -\int_x^1 \frac{1}{t}\, dt < 0 \right).$$

(iv)   If $0 < x_1 < x_2$ then $\ln x_1 < \ln x_2$.

We already know that the logarithm satisfies the multiplicative property. By applying this property repeatedly, we obtain that: If $x > 0$ and n is any integer then

$$\ln(x^n) = n \cdot \ln x.$$

A companion result is the division rule: If a and b are positive numbers then

$$\ln\left(\frac{a}{b}\right) = \ln a - \ln b.$$

EXAMPLE 6.1

Simplify the expression

$$A = \ln\left(\frac{a^3 \cdot b^2}{c^{-4} \cdot d}\right).$$

SOLUTION

We can write A in simpler terms by using the multiplicative and quotient properties:

$$\begin{aligned} A &= \ln(a^3 \cdot b^2) - \ln(c^{-4} \cdot d) \\ &= [\ln a^3 + \ln(b^2)] - [\ln(c^{-4}) + \ln d] \\ &= [3\ln a + 2 \cdot \ln b] - [(-4) \cdot \ln c + \ln d] \\ &= 3\ln a + 2 \cdot \ln b + 4 \cdot \ln c - \ln d. \end{aligned}$$

The last basic property of the logarithm is the reciprocal law: For any $x > 0$ we have

$$\ln(1/x) = -\ln x.$$

EXAMPLE 6.2

Express $\ln(1/7)$ in terms of $\ln 7$. Express $\ln(9/5)$ in terms of $\ln 3$ and $\ln 5$.

SOLUTION

We calculate that

$$\ln(1/7) = -\ln 7,$$
$$\ln(9/5) = \ln 9 - \ln 5 = \ln 3^2 - \ln 5 = 2\ln 3 - \ln 5.$$

You Try It: Simplify $\ln(a^2 b^{-3}/c^5)$.

6.1.2   THE LOGARITHM FUNCTION AND THE DERIVATIVE

Now you will see why our new definition of logarithm is so convenient. If we want to differentiate the logarithm function we can apply the Fundamental Theorem of Calculus:

$$\frac{d}{dx}\ln x = \frac{d}{dx}\int_1^x \frac{1}{t}\,dt = \frac{1}{x}.$$

More generally,

$$\frac{d}{dx}\ln u = \frac{1}{u}\frac{du}{dx}.$$

EXAMPLE 6.3

Calculate

$$\frac{d}{dx}\ln(4+x), \quad \frac{d}{dx}\ln(x^3-x), \quad \frac{d}{dx}\ln(\cos x), \quad \frac{d}{dx}[(\ln x)^5], \quad \frac{d}{dx}[(\ln x) \cdot (\cot x)].$$

SOLUTION

For the first problem, we let $u = 4 + x$ and $du/dx = 1$. Therefore we have

$$\frac{d}{dx}\ln(4+x) = \frac{1}{4+x} \cdot \frac{d}{dx}(4+x) = \frac{1}{4+x}.$$

Similarly,

$$\frac{d}{dx}\ln(x^3 - x) = \frac{1}{x^3-x} \cdot \frac{d}{dx}(x^3-x) = \frac{3x^2-1}{x^3-x}$$

$$\frac{d}{dx}\ln(\cos x) = \frac{1}{\cos x} \cdot \frac{d}{dx}(\cos x) = \frac{-\sin x}{\cos x}$$

$$\frac{d}{dx}[(\ln x)^5] = 5(\ln x)^4 \cdot \frac{d}{dx}(\ln x) = 5(\ln x)^4 \cdot \frac{1}{x} = \frac{5(\ln x)^4}{x}$$

$$\frac{d}{dx}[(\ln x)\cdot(\cot x)] = \left[\frac{d}{dx}\ln x\right]\cdot(\cot x) + (\ln x)\cdot\left[\frac{d}{dx}\cot x\right]$$

$$= \frac{1}{x}\cdot\cot x + (\ln x)\cdot(-\csc^2 x).$$

You Try It: What is the derivative of the function $\ln(x^3 + x^2)$?

Now we examine the graph of $y = \ln x$. Since

(i) $\dfrac{d}{dx}(\ln x) = \dfrac{1}{x} > 0,$

(ii) $\dfrac{d^2}{dx}(\ln x) = \dfrac{d}{dx}\left(\dfrac{1}{x}\right) = -\dfrac{1}{x^2} < 0,$

(iii) $\ln(1) = 0,$

we know that $\ln x$ is an increasing, concave down function whose graph passes through $(1, 0)$. There are no relative maxima or minima (since the derivative is never 0). Certainly $\ln .9 < 0$; the formula $\ln(.9^n) = n\ln .9$ therefore tells us that $\ln x$ is negative without bound as $x \to 0^+$. Since $\ln x = -\ln(1/x)$, we may also conclude that $\ln x$ is positive without bound as $x \to +\infty$. A sketch of the graph of $y = \ln x$ appears in Fig. 6.2.

We learned in the last paragraph that the function $\ln x$ takes negative values which are arbitrarily large in absolute value when x is small and positive. In particular, the negative y axis is a vertical asymptote. Since $\ln(1/x) = -\ln x$, we then find that $\ln x$ takes arbitrarily large positive values when x is large and positive. The graph exhibits these features.

Since we have only defined the function $\ln x$ when $x > 0$, the graph is only sketched in Fig. 6.2 to the right of the y-axis. However it certainly makes sense to

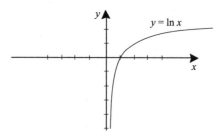

Fig. 6.2

discuss the function $\ln |x|$ when $x \neq 0$ (Fig. 6.3):

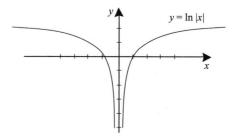

Fig. 6.3

If $x \neq 0$ then

$$\frac{d}{dx}(\ln |x|) = \frac{1}{x}.$$

In other words,

$$\int \frac{1}{x}\, dx = \ln |x| + C.$$

More generally, we have

$$\frac{d}{dx} \ln |u| = \frac{1}{u}\frac{du}{dx}$$

and

$$\int \frac{1}{u}\frac{du}{dx}\, dx = \ln |u| + C.$$

EXAMPLE 6.4

Calculate

$$\int \frac{4}{x+1}\, dx, \qquad \int \frac{1}{-2+3x}\, dx.$$

SOLUTION

$$\int \frac{4}{x+1}\,dx = 4\int \frac{1}{x+1}\,dx = 4\ln|x+1| + C$$

$$\int \frac{1}{-2+3x}\,dx = \frac{1}{3}\ln|-2+3x| + C.$$

You Try It: Calculate the integral

$$\int \frac{\cos x}{2 + \sin x}\,dx.$$

You Try It: Calculate the integral

$$\int_1^e \frac{1}{x \cdot [\ln x]^{3/2}}\,dx.$$

EXAMPLE 6.5

Evaluate the integral

$$\int \frac{\cos x}{3 \sin x - 4}\,dx.$$

SOLUTION
   For clarity we set $\varphi(x) = 3\sin x - 4$, $\varphi'(x) = 3(\cos x)$. The integral then has the form

$$\frac{1}{3}\int \frac{\varphi'(x)}{\varphi(x)}\,dx = \frac{1}{3}\ln|\varphi(x)| + C.$$

Resubstituting the expression for $\varphi(x)$ yields that

$$\int \frac{\cos x}{3\sin x - 4}\,dx = \frac{1}{3}\ln|3\sin x - 4| + C.$$

You Try It: Evaluate $\displaystyle\int \frac{x^2}{1 - x^3}\,dx.$

EXAMPLE 6.6

Calculate

$$\int \cot x\, dx.$$

SOLUTION
   We rewrite the integral as

$$\int \frac{\cos x}{\sin x}\,dx.$$

For clarity we take $\varphi(x) = \sin x$, $\varphi'(x) = \cos x$. Then the integral becomes

$$\int \frac{\varphi'(x)}{\varphi(x)} \, dx = \ln |\varphi(x)| + C.$$

Resubstituting the expression for φ yields the solution:

$$\int \cot x \, dx = \ln |\sin x| + C.$$

6.2 Exponential Basics

Examine Fig. 6.4, which shows the graph of the function

$$f(x) = \ln x, \quad x > 0.$$

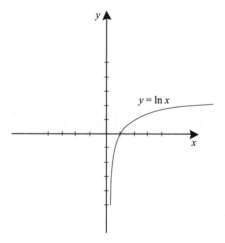

$y = \ln x$

Fig. 6.4

As we observed in Section 1, the function f takes on all real values. We already have noticed that, since

$$\frac{d}{dx} \ln x = \frac{1}{x} > 0,$$

the function $\ln x$ is increasing. As a result,

$$\ln : \{x : x > 0\} \to \mathbb{R}$$

is one-to-one and onto. Hence the natural logarithm function has an inverse.

The inverse function to the natural logarithm function is called the *exponential function* and is written $\exp(x)$. The domain of exp is the entire real line. The range is the set of positive real numbers.

EXAMPLE 6.7

Using the definition of the exponential function, simplify the expressions

$$\exp(\ln a + \ln b) \quad \text{and} \quad \ln(7 \cdot [\exp(c)]).$$

SOLUTION

We use the key property that the exponential function is the inverse of the logarithm function. We have

$$\exp(\ln a + \ln b) = \exp(\ln(a \cdot b)) = a \cdot b,$$
$$\ln(7 \cdot [\exp(c)]) = \ln 7 + \ln(\exp(c)) = \ln 7 + c.$$

You Try It: Simplify the expression $\ln(a^3 \cdot 3^5 \cdot 5^{-4})$.

6.2.1   FACTS ABOUT THE EXPONENTIAL FUNCTION

First review the properties of inverse functions that we learned in Subsection 1.8.5. The graph of $\exp(x)$ is obtained by reflecting the graph of $\ln x$ in the line $y = x$. We exhibit the graph of $y = \exp(x)$ in Fig. 6.5.

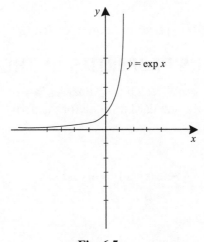

Fig. 6.5

We see, from inspection of this figure, that $\exp(x)$ is increasing and is concave up. Since $\ln(1) = 0$ we may conclude that $\exp(0) = 1$. Next we turn to some of the algebraic properties of the exponential function.

For all real numbers a and b we have

(a) $\exp(a + b) = [\exp(a)] \cdot [\exp(b)]$.

(b) For any a and b we have $\exp(a - b) = \dfrac{\exp(a)}{\exp(b)}$.

These properties are verified just by exploiting the fact that the exponential is the inverse of the logarithm, as we saw in Example 6.7.

EXAMPLE 6.8

Use the basic properties to simplify the expression

$$\frac{[\exp(a)]^2 \cdot [\exp(b)]^3}{[\exp(c)]^4}.$$

SOLUTION

We calculate that

$$\frac{[\exp(a)]^2 \cdot [\exp(b)]^3}{[\exp(c)]^4} = \frac{[\exp(a)] \cdot [\exp(a)] \cdot [\exp(b)] \cdot [\exp(b)] \cdot [\exp(b)]}{[\exp(c)] \cdot [\exp(c)] \cdot [\exp(c)] \cdot [\exp(c)]}$$

$$= \frac{\exp(a + a + b + b + b)}{\exp(c + c + c + c)}$$

$$= \exp(a + a + b + b + b - c - c - c - c)$$

$$= \exp(2a + 3b - 4c).$$

You Try It: Simplify the expression $(\exp a)^{-3} \cdot (\exp b)^2 / \exp(5c)$.

6.2.2 CALCULUS PROPERTIES OF THE EXPONENTIAL

Now we want to learn some "calculus properties" of our new function $\exp(x)$. These are derived from the standard formula for the derivative of an inverse, as in Section 2.5.1.

For all x we have

$$\frac{d}{dx}(\exp(x)) = \exp(x).$$

In other words,

$$\int \exp(x)\, dx = \exp(x).$$

More generally,

$$\frac{d}{dx}\exp(u) = \exp(u)\frac{du}{dx}$$

and

$$\int \exp(u) \frac{du}{dx} \, dx = \exp(u) + C.$$

We note for the record that the exponential function is the *only function* (up to constant multiples) that is its own derivative. This fact will come up later in our applications of the exponential

EXAMPLE 6.9

Compute the derivatives:

$$\frac{d}{dx} \exp(4x), \quad \frac{d}{dx}(\exp(\cos x)), \quad \frac{d}{dx}([\exp(x)] \cdot [\cot x]).$$

SOLUTION

For the first problem, notice that $u = 4x$ hence $du/dx = 4$. Therefore we have

$$\frac{d}{dx} \exp(4x) = [\exp(4x)] \cdot \frac{d}{dx}(4x) = 4 \cdot \exp(4x).$$

Similarly,

$$\frac{d}{dx}(\exp(\cos x)) = [\exp(\cos x)] \cdot \left(\frac{d}{dx} \cos x \right) = [\exp(\cos x)] \cdot (-\sin x),$$

$$\frac{d}{dx}([\exp(x)] \cdot [\cot x]) = \left[\frac{d}{dx} \exp(x) \right] \cdot (\cot x) + [\exp(x)] \cdot \left(\frac{d}{dx} \cot x \right)$$

$$= [\exp(x)] \cdot (\cot x) + [\exp(x)] \cdot (-\csc^2 x).$$

You Try It: Calculate $(d/dx)(\exp(x \cdot \sin x))$.

EXAMPLE 6.10

Calculate the integrals:

$$\int \exp(5x) \, dx, \quad \int [\exp(x)]^3 \, dx, \quad \int \exp(2x + 7) \, dx.$$

SOLUTION

We have

$$\int \exp(5x) \, dx = \frac{1}{5} \exp(5x) + C$$

$$\int [\exp(x)]^3 \, dx = \int [\exp(x)] \cdot [\exp(x)] \cdot [\exp(x)] \, dx$$

$$= \int \exp(3x) \, dx = \frac{1}{3} \exp(3x) + C$$

$$\int \exp(2x + 7) \, dx = \frac{1}{2} \int \exp(2x + 7) \cdot 2 \, dx = \frac{1}{2} \exp(2x + 7) + C.$$

EXAMPLE 6.11

Evaluate the integral

$$\int [\exp(\cos^3 x)] \cdot \sin x \cdot \cos^2 x \, dx.$$

SOLUTION

For clarity, we let $\varphi(x) = \cos^3 x$, $\varphi'(x) = 3\cos^2 x \cdot (-\sin x)$. Then the integral becomes

$$-\frac{1}{3} \int \exp(\varphi(x)) \cdot \varphi'(x) \, dx = -\frac{1}{3} \exp(\varphi(x)) + C.$$

Resubstituting the expression for $\varphi(x)$ then gives

$$\int [\exp(\cos^3 x)] \cdot \sin x \cdot \cos^2 x \, dx = -\frac{1}{3} \exp(\cos^3 x) + C.$$

EXAMPLE 6.12

Evaluate the integral

$$\int \frac{\exp(x) + \exp(-x)}{\exp(x) - \exp(-x)} \, dx.$$

SOLUTION

For clarity, we set $\varphi(x) = \exp(x) - \exp(-x)$, $\varphi'(x) = \exp(x) + \exp(-x)$. Then our integral becomes

$$\int \frac{\varphi'(x) \, dx}{\varphi(x)} = \ln |\varphi(x)| + C.$$

Resubstituting the expression for $\varphi(x)$ gives

$$\int \frac{\exp(x) + \exp(-x)}{\exp(x) - \exp(-x)} \, dx = \ln |\exp(x) - \exp(-x)| + C.$$

You Try It: Calculate $\int x \cdot \exp(x^2 - 3) \, dx$.

6.2.3 THE NUMBER e

The number $\exp(1)$ is a special constant which arises in many mathematical and physical contexts. It is denoted by the symbol e in honor of the Swiss mathematician Leonhard Euler (1707–1783) who first studied this constant. We next see how to calculate the decimal expansion for the number e.

In fact, as can be proved in a more advanced course, Euler's constant e satisfies the identity

$$\lim_{n \to +\infty} \left(1 + \frac{1}{n}\right)^n = e.$$

[Refer to the "You Try It" following Example 5.9 in Subsection 5.2.3 for a consideration of this limit.]

This formula tells us that, for large values of n, the expression

$$\left(1 + \frac{1}{n}\right)^n$$

gives a good approximation to the value of e. Use your calculator or computer to check that the following calculations are correct:

$$n = 10 \qquad \left(1 + \frac{1}{n}\right)^n = 2.5937424601$$

$$n = 50 \qquad \left(1 + \frac{1}{n}\right)^n = 2.69158802907$$

$$n = 100 \qquad \left(1 + \frac{1}{n}\right)^n = 2.70481382942$$

$$n = 1000 \qquad \left(1 + \frac{1}{n}\right)^n = 2.71692393224$$

$$n = 10000000 \qquad \left(1 + \frac{1}{n}\right)^n = 2.71828169254.$$

With the use of a sufficiently large value of n, together with estimates for the error term

$$\left| e - \left(1 + \frac{1}{n}\right)^n \right|,$$

it can be determined that

$$e = 2.71828182846$$

to eleven place decimal accuracy. Like the number π, the number e is an irrational number. Notice that, since $\exp(1) = e$, we also know that $\ln e = 1$.

EXAMPLE 6.13

Simplify the expression

$$\ln(e^5 \cdot 8^{-3}).$$

SOLUTION

We calculate that

$$\ln(e^5 \cdot 8^{-3}) = \ln(e^5) + \ln(8^{-3})$$
$$= 5\ln(e) - 3\ln 8$$
$$= 5 - 3\ln 8.$$

You Try It: Use your calculator to compute $\log_{10} e$ and $\ln 10 = \log_e 10$ (see Example 6.20 below). Confirm that these numbers are reciprocals of each other.

6.3 Exponentials with Arbitrary Bases

6.3.1 ARBITRARY POWERS

We know how to define integer powers of real numbers. For instance

$$6^4 = 6 \cdot 6 \cdot 6 \cdot 6 = 1296 \quad \text{and} \quad 9^{-3} = \frac{1}{9 \cdot 9 \cdot 9} = \frac{1}{729}.$$

But what does it mean to calculate

$$4^\pi \quad \text{or} \quad \pi^e?$$

You can calculate values for these numbers by punching suitable buttons on your calculator, but that does not explain what the numbers mean or how the calculator was programmed to calculate them. We will use our understanding of the exponential and logarithm functions to now define these exponential expressions.

If $a > 0$ and b is any real number then we define

$$a^b = \exp(b \cdot \ln a). \tag{$*$}$$

To come to grips with this rather abstract formulation, we begin to examine some properties of this new notion of exponentiation:

If a is a positive number and b is any real number then

(1) $\ln(a^b) = b \cdot \ln a.$

In fact

$$\ln(a^b) = \ln(\exp(b \cdot \ln a)).$$

But ln and exp are inverse, so that the last expression simplifies to $b \cdot \ln a$.

EXAMPLE 6.14

Let $a > 0$. Compare the new definition of a^4 with the more elementary definition of a^4 in terms of multiplying a by itself four times.

SOLUTION

We ordinarily think of a^4 as meaning

$$a \cdot a \cdot a \cdot a.$$

According to our new definition of a^b we have

$$a^4 = \exp(4 \cdot \ln a) = \exp(\ln a + \ln a + \ln a + \ln a)$$
$$= \exp(\ln[a \cdot a \cdot a \cdot a]) = a \cdot a \cdot a \cdot a.$$

It is reassuring to see that our new definition of exponentiation is consistent with the familiar notion for integer exponents.

EXAMPLE 6.15

Express exp(x) as a power of e.

SOLUTION

According to our definition,

$$e^x = \exp(x \cdot \ln(e)).$$

But we learned in the last section that $\ln(e) = 1$. As a result,

$$e^x = \exp(x).$$

You Try It: Simplify the expression $\ln[e^x \cdot x^e]$.

Because of this last example we will not in the future write the exponential function as exp(x) but will use the more common notation e^x. Thus

$\exp(\ln x) = x$	becomes	$e^{\ln x} = x$
$\ln(\exp(x)) = x$	becomes	$\ln(e^x) = x$
$\exp(a + b) = [\exp(a)] \cdot [\exp(b)]$	becomes	$e^{a+b} = e^a e^b$
$\exp(a - b) = \dfrac{\exp(a)}{\exp(b)}$	becomes	$e^{a-b} = \dfrac{e^a}{e^b}$
$a^b = \exp(b \cdot \ln a)$	becomes	$a^b = e^{b \cdot \ln a}$.

EXAMPLE 6.16

Use our new definitions to simplify the expression

$$A = e^{[5 \cdot \ln 2 - 3 \cdot \ln 4]}.$$

SOLUTION

We write

$$A = e^{[\ln(2^5) - \ln(4^3)]} = e^{\ln 32 - \ln 64} = \frac{e^{\ln(32)}}{e^{\ln(64)}} = \frac{32}{64} = \frac{1}{2}.$$

We next see that our new notion of exponentiation satisfies certain familiar rules. If $a, d > 0$ and $b, c \in \mathbb{R}$ then

(i) $a^{b+c} = a^b \cdot a^c$

(ii) $a^{b-c} = \dfrac{a^b}{a^c}$

(iii) $(a^b)^c = a^{b \cdot c}$

(iv) $a^b = d$ if and only if $d^{1/b} = a$ (provided $b \neq 0$)

(v) $a^0 = 1$

(vi) $a^1 = a$

(vii) $(a \cdot d)^c = a^c \cdot d^c$.

EXAMPLE 6.17

Simplify each of the expressions

$$(e^4)^{\ln 3}, \quad \frac{5^{-7} \cdot \pi^4}{5^{-3} \cdot \pi^2}, \quad (3^2 \cdot x^3)^4.$$

SOLUTION

We calculate:

$$(e^4)^{\ln 3} = e^{4 \cdot \ln 3} = (e^{\ln 3})^4 = 3^4 = 81;$$

$$\frac{5^{-7} \cdot \pi^4}{5^{-3} \cdot \pi^2} = 5^{-7-(-3)} \cdot \pi^{4-2} = 5^{-4} \cdot \pi^2 = \frac{1}{625} \cdot \pi^2;$$

$$(3^2 \cdot x^3)^4 = (3^2)^4 \cdot (x^3)^4 = 3^8 \cdot x^{12} = 6561 \cdot x^{12}.$$

You Try It: Simplify the expression $\ln[e^{3x} \cdot e^{-y-5} \cdot 2^4]$.

EXAMPLE 6.18

Solve the equation

$$(x^3 \cdot 5)^8 = 9$$

for *x*.

SOLUTION

We have

$$(x^3 \cdot 5)^8 = 9 \implies x^3 \cdot 5 = 9^{1/8} \implies x^3 = 9^{1/8} \cdot 5^{-1}$$

$$\implies x = (9^{1/8} \cdot 5^{-1})^{1/3} \implies x = \frac{9^{1/24}}{5^{1/3}}.$$

You Try It: Solve the equation $4^x \cdot 3^{2x} = 7$. [*Hint*: Take the logarithm of both sides. See also Example 6.22 below.]

6.3.2 LOGARITHMS WITH ARBITRARY BASES

If you review the first few paragraphs of Section 1, you will find an intuitively appealing definition of the logarithm to the base 2:

$\log_2 x$ is the power to which you need to raise 2 to obtain x.

With this intuitive notion we readily see that

$\log_2 16 =$ "the power to which we raise 2 to obtain 16" $= 4$

and

$\log_2(1/4) =$ "the power to which we raise 2 to obtain 1/4" $= -2.$

However this intuitive approach does not work so well if we want to take $\log_\pi 5$ or $\log_2 \sqrt{7}$. Therefore we will give a new definition of the logarithm to any base $a > 0$ which in simple cases coincides with the intuitive notion of logarithm.

If $a > 0$ and $b > 0$ then

$$\log_a b = \frac{\ln b}{\ln a}.$$

EXAMPLE 6.19

Calculate $\log_2 32$.

SOLUTION

We see that

$$\log_2 32 = \frac{\ln 32}{\ln 2} = \frac{\ln 2^5}{\ln 2} = \frac{5 \cdot \ln 2}{\ln 2} = 5.$$

Notice that, in this example, the new definition of $\log_2 32$ agrees with the intuitive notion just discussed.

EXAMPLE 6.20

Express $\ln x$ as the logarithm to some base.

SOLUTION

If $x > 0$ then

$$\log_e x = \frac{\ln x}{\ln e} = \frac{\ln x}{1} = \ln x.$$

Thus we see that the natural logarithm $\ln x$ is precisely the same as $\log_e x$.

Math Note: In mathematics, it is common to write $\ln x$ rather than $\log_e x$.

You Try It: Calculate $\log_3 27 + \log_5(1/25) - \log_2 8$.

We will be able to do calculations much more easily if we learn some simple properties of logarithms and exponentials.

If $a > 0$ and $b > 0$ then

$$a^{(\log_a b)} = b.$$

If $a > 0$ and $b \in \mathbb{R}$ is arbitrary then

$$\log_a(a^b) = b.$$

If $a > 0$, $b > 0$, and $c > 0$ then

  (i) $\log_a(b \cdot c) = \log_a b + \log_a c$

  (ii) $\log_a(b/c) = \log_a b - \log_a c$

  (iii) $\log_a b = \dfrac{\log_c b}{\log_c a}$

  (iv) $\log_a b = \dfrac{1}{\log_b a}$

  (v) $\log_a 1 = 0$

  (vi) $\log_a a = 1$

  (vii)   For any exponent α, $\log_a(b^\alpha) = \alpha \cdot (\log_a b)$

We next give several examples to familiarize you with logarithmic and exponential operations.

EXAMPLE 6.21

Simplify the expression

$$\log_3 81 - 5 \cdot \log_2 8 - 3 \cdot \ln(e^4).$$

SOLUTION

The expression equals

$$\log_3(3^4) - 5 \cdot \log_2(2^3) - 3 \cdot \ln e^4 = 4 \cdot \log_3 3 - 5 \cdot [3 \cdot \log_2 2] - 3 \cdot [4 \cdot \ln e]$$
$$= 4 \cdot 1 - 5 \cdot 3 \cdot 1 - 3 \cdot 4 \cdot 1 = -23.$$

You Try It: What does $\log_3 5$ mean in terms of natural logarithms?

EXAMPLE 6.22

Solve the equation

$$5^x \cdot 2^{3x} = \frac{4}{7^x}$$

for the unknown x.

SOLUTION

We take the natural logarithm of both sides:

$$\ln(5^x \cdot 2^{3x}) = \ln\left(\frac{4}{7^x}\right).$$

Applying the rules for logarithms we obtain

$$\ln(5^x) + \ln(2^{3x}) = \ln 4 - \ln(7^x)$$

or

$$x \cdot \ln 5 + 3x \cdot \ln 2 = \ln 4 - x \cdot \ln 7.$$

Gathering together all the terms involving x yields

$$x \cdot [\ln 5 + 3 \cdot \ln 2 + \ln 7] = \ln 4$$

or

$$x \cdot [\ln(5 \cdot 2^3 \cdot 7)] = \ln 4.$$

Solving for x gives

$$x = \frac{\ln 4}{\ln 280} = \log_{280} 4.$$

EXAMPLE 6.23

Simplify the expression

$$B = \frac{5 \cdot \log_7 3 - (1/4) \cdot \log_7 16}{3 \cdot \log_7 5 + (1/5) \cdot \log_7 32}.$$

SOLUTION

The numerator of B equals

$$\log_7(3^5) - \log_7(16^{1/4}) = \log_7 243 - \log_7 2 = \log_7(243/2).$$

Similarly, the denominator can be rewritten as

$$\log_7 5^3 + \log_7(32^{1/5}) = \log_7 125 + \log_7 2 = \log_7(125 \cdot 2) = \log_7 250.$$

Putting these two results together, we find that

$$B = \frac{\log_7 243/2}{\log_7 250} = \log_{250}(243/2).$$

You Try It: What does $3^{\sqrt{2}}$ mean (in terms of the natural logarithm function)?

EXAMPLE 6.24

Simplify the expression $(\log_4 9) \cdot (\log_9 16)$.

SOLUTION

We have

$$(\log_4 9) \cdot (\log_9 15) = \left(\frac{1}{\log_9 4}\right) \cdot \log_9 16 = \log_4 16 = 2.$$

6.4 Calculus with Logs and Exponentials to Arbitrary Bases

6.4.1 DIFFERENTIATION AND INTEGRATION OF $\log_a x$ AND a^x

We begin by noting these facts:

If $a > 0$ then

(i) $\dfrac{d}{dx}a^x = a^x \cdot \ln a$; equivalently, $\displaystyle\int a^x\,dx = \dfrac{a^x}{\ln a} + C.$

(ii) $\dfrac{d}{dx}(\log_a x) = \dfrac{1}{x \cdot \ln a}$

Math Note: As always, we can state these last formulas more generally as

$$\frac{d}{dx}a^u = a^u \cdot \frac{du}{dx} \cdot \ln a$$

and

$$\frac{d}{dx}\log_a u = \frac{1}{u} \cdot \frac{du}{dx} \cdot \frac{1}{\ln a}.$$

EXAMPLE 6.25

Calculate

$$\frac{d}{dx}(5^x), \quad \frac{d}{dx}(3^{\cos x}), \quad \frac{d}{dx}(\log_8 x), \quad \frac{d}{dx}(\log_4(x \cdot \cos x)).$$

SOLUTION

We see that

$$\frac{d}{dx}(5^x) = 5^x \cdot \ln 5.$$

For the second problem, we apply our general formulation with $a = 3$, $u = \cos x$ to obtain

$$\frac{d}{dx}(3^{\cos x}) = 3^{\cos x} \cdot \left(\frac{d}{dx}\cos x\right) \cdot \ln 3 = 3^{\cos x} \cdot (-\sin x) \cdot \ln 3.$$

Similarly,

$$\frac{d}{dx}(\log_8 x) = \frac{1}{x \cdot \ln 8}$$

$$\frac{d}{dx}(\log_4(x \cdot \cos x)) = \frac{1}{(x \cdot \cos x) \cdot \ln 4} \cdot \frac{d}{dx}(x \cdot \cos x)$$

$$= \frac{\cos x + (x \cdot (-\sin x))}{(x \cdot \cos x) \cdot \ln 4}.$$

EXAMPLE 6.26

Integrate

$$\int 3^{\cot x} \cdot (-\csc^2 x) \, dx.$$

SOLUTION

For clarity we set $\varphi(x) = \cot x, \varphi'(x) = -\csc^2 x$. Then our integral becomes

$$\int 3^{\varphi(x)} \cdot \varphi'(x) \, dx = \left(\frac{1}{\ln 3}\right) \cdot \int 3^{\varphi(x)} \cdot \varphi'(x) \cdot \ln 3 \, dx$$

$$= \left(\frac{1}{\ln 3}\right) \cdot 3^{\varphi(x)} + C.$$

Resubstituting the expression for $\varphi(x)$ now gives that

$$\int 3^{\cot x} \cdot (-\csc^2 x) \, dx = \frac{1}{\ln 3} \cdot 3^{\cot x} + C.$$

You Try It: Evaluate $\int (\log_6(x^3)/x) \, dx$.

You Try It: Calculate the integral

$$\int x \cdot 3^{x^2} \, dx.$$

Our new ideas about arbitrary exponents and bases now allow us to formulate a general result about derivatives of powers:

For any real exponent a we have

$$\frac{d}{dx}x^a = a \cdot x^{a-1}.$$

EXAMPLE 6.27

Calculate the derivative of $x^{-\pi}$, $x^{\sqrt{3}}$, x^e.

SOLUTION
We have

$$\frac{d}{dx}x^{-\pi} = -\pi \cdot x^{-\pi-1},$$

$$\frac{d}{dx}x^{\sqrt{3}} = \sqrt{3} \cdot x^{\sqrt{3}-1},$$

$$\frac{d}{dx}x^{e} = e \cdot x^{e-1}.$$

You Try It: Calculate $(d/dx)5^{\sin x - x^2}$. Calculate $(d/dx)x^{4\pi}$.

6.4.2   GRAPHING OF LOGARITHMIC AND EXPONENTIAL FUNCTIONS

If $a > 0$ and $f(x) = \log_a x$, $x > 0$, then

$$f'(x) = \frac{1}{x \cdot \ln a}$$

$$f''(x) = \frac{-1}{x^2 \cdot \ln a}$$

$$f(1) = 0.$$

Using this information, we can sketch the graph of $f(x) = \log_a x$.

If $a > 1$ then $\ln a > 0$ so that $f'(x) > 0$ and $f''(x) < 0$. The graph of f is exhibited in Fig. 6.6.

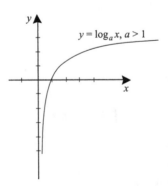

Fig. 6.6

If $0 < a < 1$ then $\ln a = -\ln(1/a) < 0$ so that $f'(x) < 0$ and $f''(x) > 0$. The graph of f is sketched in Fig. 6.7.

Since $g(x) = a^x$ is the inverse function to $f(x) = \log_a x$, the graph of g is the reflection in the line $y = x$ of the graph of f (Figs 6.6 and 6.7). See Figs 6.8, 6.9.

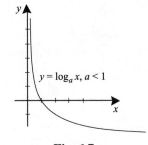

Fig. 6.7

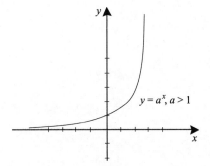

Fig. 6.8

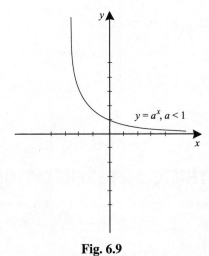

Fig. 6.9

Figure 6.10 shows the graphs of $\log_a x$ for several different values of $a > 1$. Figure 6.11 shows the graphs of a^x for several different values of $a > 1$.

You Try It: Sketch the graph of $y = 4^x$ and $y = \log_4 x$ on the same set of axes.

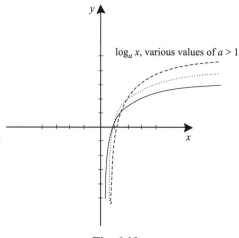

Fig. 6.10

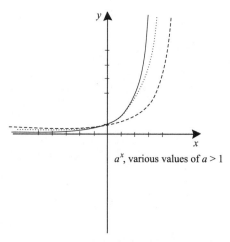

Fig. 6.11

6.4.3   LOGARITHMIC DIFFERENTIATION

We next show how to use the logarithm as an aid to differentiation. The key idea is that if F is a function taking positive values then we can exploit the formula

$$[\ln F]' = \frac{F'}{F}. \qquad (*)$$

EXAMPLE 6.28

Calculate the derivative of the function

$$F(x) = (\cos x)^{(\sin x)}, \quad 0 < x < \pi.$$

SOLUTION

We take the natural logarithm of both sides:

$$\ln F(x) = \ln\big((\cos x)^{(\sin x)}\big) = (\sin x) \cdot (\ln(\cos x)). \qquad (\dagger)$$

Now we calculate the derivative using the formula (∗) preceding this example: The derivative of the left side of (†) is

$$\frac{F'(x)}{F(x)}.$$

Using the product rule, we see that the derivative of the far right side of (†) is

$$(\cos x) \cdot (\ln(\cos x)) + (\sin x) \cdot \left(\frac{-\sin x}{\cos x}\right).$$

We conclude that

$$\frac{F'(x)}{F(x)} = (\cos x) \cdot (\ln(\cos x)) + (\sin x) \cdot \left(\frac{-\sin x}{\cos x}\right).$$

Thus

$$F'(x) = \left[(\cos x) \cdot (\ln(\cos x)) - \frac{\sin^2 x}{\cos x}\right] \cdot F(x)$$

$$= \left[(\cos x) \cdot \ln(\cos x) - \frac{\sin^2 x}{\cos x}\right] \cdot (\cos x)^{(\sin x)}$$

You Try It: Differentiate $\log_9 |\cos x|$.

You Try It: Differentiate $3^{\sin(3x)}$. Differentiate $x^{\sin 3x}$.

EXAMPLE 6.29

Calculate the derivative of $F(x) = x^2 \cdot (\sin x) \cdot 5^x$.

SOLUTION

We have

$$[\ln F(x)]' = [\ln(x^2 \cdot (\sin x) \cdot 5^x)]'$$

$$= [(2 \cdot \ln x) + \ln(\sin x) + (x \cdot \ln 5)]'$$

$$= \frac{2}{x} + \frac{\cos x}{\sin x} + \ln 5.$$

Using formula (∗), we conclude that

$$\frac{F'(x)}{F(x)} = \frac{2}{x} + \frac{\cos x}{\sin x} + \ln 5$$

hence

$$F'(x) = \left[\frac{2}{x} + \frac{\cos x}{\sin x} + \ln 5\right] \cdot [x^2 \cdot (\sin x) \cdot 5^x].$$

You Try It: Calculate $(d/dx)[(\ln x)^{\ln x}]$.

6.5   Exponential Growth and Decay

Many processes of nature and many mathematical applications involve logarithmic and exponential functions. For example, if we examine a population of bacteria, we notice that the rate at which the population grows is proportional to the number of bacteria present. To see that this makes good sense, suppose that a bacterium reproduces itself every 4 hours. If we begin with 5 thousand bacteria then

after 4 hours	there are	10 thousand bacteria
after 8 hours	there are	20 thousand bacteria
after 12 hours	there are	40 thousand bacteria
after 16 hours	there are	80 thousand bacteria ...
etc.		

The point is that each new generation of bacteria *also reproduces*, and the older generations reproduce as well. A sketch (Fig. 6.12) of the bacteria population against time shows that the growth is certainly not linear—indeed the shape of the curve appears to be of exponential form.

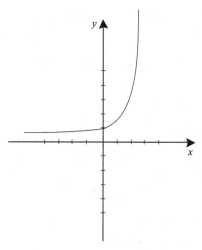

Fig. 6.12

Notice that when the number of bacteria is large, then different generations of bacteria will be reproducing at different times. So, averaging out, it makes sense to hypothesize that the growth of the bacteria population varies continuously as in Fig. 6.13. Here we are using a standard device of mathematical analysis: even though the number of bacteria is always an integer, we represent the graph of the population of bacteria by a smooth curve. This enables us to apply the tools of calculus to the problem.

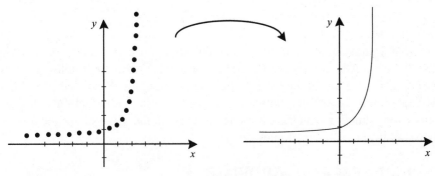

Fig. 6.13

6.5.1   A DIFFERENTIAL EQUATION

If $B(t)$ represents the number of bacteria present in a given population at time t, then the preceding discussion suggests that

$$\frac{dB}{dt} = K \cdot B(t),$$

where K is a constant of proportionality. This equation expresses quantitatively the assertion that the rate of change of $B(t)$ (that is to say, the quantity dB/dt) is proportional to $B(t)$. To solve this equation, we rewrite it as

$$\frac{1}{B(t)} \cdot \frac{dB}{dt} = K.$$

We integrate both sides with respect to the variable t:

$$\int \frac{1}{B(t)} \cdot \frac{dB}{dt} \, dt = \int K \, dt.$$

The left side is

$$\ln |B(t)| + C$$

and the right side is

$$Kt + \tilde{C},$$

where C and $\widetilde{C}$ are constants of integration. We thus obtain

$$\ln |B(t)| = Kt + D,$$

where we have amalgamated the two constants into a single constant D. Exponentiating both sides gives

$$|B(t)| = e^{Kt+D}$$

or

$$B(t) = e^D \cdot e^{Kt} = P \cdot e^{Kt}. \qquad (\star)$$

Notice that we have omitted the absolute value signs since the number of bacteria is always positive. Also we have renamed the constant e^D with the simpler symbol P.

Equation $(\star)$ will be our key to solving exponential growth and decay problems.

We motivated our calculation by discussing bacteria, but in fact the calculation applies to any function which grows at a rate proportional to the size of the function.

Next we turn to some examples.

6.5.2   BACTERIAL GROWTH

EXAMPLE 6.30

A population of bacteria tends to double every four hours. If there are 5000 bacteria at 9:00 a.m., then how many will there be at noon?

SOLUTION

To answer this question, let $B(t)$ be the number of bacteria at time t. For convenience, let $t = 0$ correspond to 9:00 a.m. and suppose that time is measured in hours. Thus noon corresponds to $t = 3$.

Equation $(\star)$ guarantees that

$$B(t) = P \cdot e^{Kt}$$

for some undetermined constants P and K. We also know that

$$5000 = B(0) = P \cdot e^{K \cdot 0} = P.$$

We see that $P = 5000$ and $B(t) = 5000 \cdot e^{Kt}$. We still need to solve for K.

Since the population tends to double in four hours, there will be 10,000 bacteria at time $t = 4$; hence

$$10000 = B(4) = 5000 \cdot e^{K \cdot 4}.$$

We divide by 5000 to obtain

$$2 = e^{K \cdot 4}.$$

Taking the natural logarithm of both sides yields

$$\ln 2 = \ln(e^{K \cdot 4}) = 4K.$$

We conclude that $K = [\ln 2]/4$. As a result,

$$B(t) = 5000 \cdot \left(e^{([\ln 2]/4)t}\right).$$

We simplify this equation by noting that

$$e^{([\ln 2]/4)t} = (e^{\ln 2})^{t/4} = 2^{t/4}.$$

In conclusion,

$$B(t) = 5000 \cdot 2^{t/4}.$$

The number of bacteria at noon (time $t = 3$) is then given by

$$B(3) = 5000 \cdot 2^{3/4} \approx 8409.$$

It is important to realize that population growth problems cannot be described using just arithmetic. Exponential growth is nonlinear, and advanced analytical ideas (such as calculus) must be used to understand it.

EXAMPLE 6.31

Suppose that a certain petri dish contains 6000 bacteria at 9:00 p.m. and 10000 bacteria at 11:00 p.m. How many of the bacteria were there at 7:00 p.m.?

SOLUTION

We know that

$$B(t) = P \cdot e^{Kt}.$$

The algebra is always simpler if we take one of the times in the initial data to correspond to $t = 0$. So let us say that 9:00 p.m. is $t = 0$. Then 11:00 p.m. is $t = 2$ and 7:00 p.m. is $t = -2$. The initial data then tell us that

$$6000 = P \cdot e^{K \cdot 0} \qquad\qquad (*)$$
$$10000 = P \cdot e^{K \cdot 2}. \qquad\qquad (**)$$

From equation $(*)$ we may immediately conclude that $P = 6000$. Substituting this into $(**)$ gives

$$10000 = 6000 \cdot (e^K)^2.$$

We conclude that

$$e^K = \frac{\sqrt{5}}{\sqrt{3}}.$$

As a result,

$$B(t) = 6000 \cdot \left(\frac{\sqrt{5}}{\sqrt{3}}\right)^t.$$

At time $t = -2$ (7:00 p.m.) the number of bacteria was therefore

$$B(-2) = 6000 \cdot \left(\frac{\sqrt{5}}{\sqrt{3}}\right)^{-2} = \frac{3}{5} \cdot 6000 = 3600.$$

You Try It: A petri dish has 5000 bacteria at 1:00 p.m. on a certain day and 8000 bacteria at 5:00 p.m. that same day. How many bacteria were there at noon?

6.5.3   RADIOACTIVE DECAY

Another natural phenomenon which fits into our theoretical framework is *radioactive decay*. Radioactive material, such as C^{14} (radioactive carbon), has a *half life*. Saying that the half life of a material is h years means that if A grams of material is present at time t then $A/2$ grams will be present at time $t + h$. In other words, half of the material decays every h years. But this is another way of saying that the rate at which the radioactive material vanishes is proportional to the amount present. So equation (⋆) will apply to problems about radioactive decay.

EXAMPLE 6.32

Five grams of a certain radioactive isotope decay to three grams in 100 years. After how many more years will there be just one gram?

SOLUTION

First note that the answer is *not* "we lose two grams every hundred years so ...." The rate of decay depends on the amount of material present. That is the key.

Instead, we let $R(t)$ denote the amount of radioactive material at time t. Equation (⋆) guarantees that R has the form

$$R(t) = P \cdot e^{Kt}.$$

Letting $t = 0$ denote the time at which there are 5 grams of isotope, and measuring time in years, we have

$$R(0) = 5 \quad \text{and} \quad R(100) = 3.$$

From the first piece of information we learn that

$$5 = P \cdot e^{K \cdot 0} = P.$$

Hence $P = 5$ and

$$R(t) = 5 \cdot e^{Kt} = 5 \cdot (e^K)^t.$$

The second piece of information yields

$$3 = R(100) = 5 \cdot (e^K)^{100}.$$

We conclude that

$$(e^K)^{100} = \frac{3}{5}$$

or

$$e^K = \left(\frac{3}{5}\right)^{1/100}.$$

Thus the formula for the amount of isotope present at time t is

$$R(t) = 5 \cdot \left(\frac{3}{5}\right)^{t/100}.$$

Thus we have complete information about the function R, and we can answer the original question.

There will be 1 gram of material present when

$$1 = R(t) = 5 \cdot \left(\frac{3}{5}\right)^{t/100}$$

or

$$\frac{1}{5} = \left(\frac{3}{5}\right)^{t/100}.$$

We solve for t by taking the natural logarithm of both sides:

$$\ln(1/5) = \ln\left[\left(\frac{3}{5}\right)^{t/100}\right] = \frac{t}{100} \cdot \ln(3/5).$$

We conclude that there is 1 gram of radioactive material remaining when

$$t = 100 \cdot \frac{\ln(1/5)}{\ln(3/5)} \approx 315.066.$$

So at time $t = 315.066$, or after 215.066 more years, there will be 1 gram of the isotope remaining.

You Try It: Our analysis of exponential growth and decay is derived from the hypothesis that the rate of growth is proportional to the amount of matter present. Suppose instead that we are studying a system in which the rate of decay is proportional to the *square* of the amount of matter. Let $M(t)$ denote the amount of matter at time t. Then our physical law is expressed as

$$\frac{dM}{dt} = C \cdot M^2.$$

Here C is a (negative) constant of proportionality. We apply the method of "separation of variables" described earlier in the section. Thus

$$\frac{dM/dt}{M^2} = C$$

so that

$$\int \frac{dM/dt}{M^2}\, dt = \int C\, dt.$$

Evaluating the integrals, we find that

$$-\frac{1}{M} = Ct + D.$$

We have combined the constants from the two integrations. In summary,

$$M(t) = -\frac{1}{Ct + D}.$$

For the problem to be realistic, we will require that $C < 0$ (so that $M > 0$ for large values of t) and we see that the population decays like the reciprocal of a linear function when t becomes large.

Re-calculate Example 6.32 using this new law of exponential decay.

6.5.4 COMPOUND INTEREST

Yet a third illustration of exponential growth is in the compounding of interest. If principal P is put in the bank at p percent simple interest per year then after one year the account has

$$P \cdot \left(1 + \frac{p}{100}\right)$$

dollars. [Here we assume, of course, that all interest is reinvested in the account.] But if the interest is *compounded* n times during the year then the year is divided into n equal pieces and at each time interval of length $1/n$ an interest payment of percent p/n is added to the account. Each time this fraction of the interest is added to the account, the money in the account is multiplied by

$$1 + \frac{p/n}{100}.$$

Since this is done n times during the year, the result at the end of the year is that the account holds

$$P \cdot \left(1 + \frac{p}{100n}\right)^n \qquad\qquad (*)$$

dollars at the end of the year. Similarly, at the end of t years, the money accumulated will be

$$P \cdot \left(1 + \frac{p}{100n}\right)^{nt}.$$

Let us set

$$k = \frac{n \cdot 100}{p}$$

and rewrite ($*$) as

$$P \cdot \left[1 + \frac{1}{k}\right]^{kp/100} = P \cdot \left[\left(1 + \frac{1}{k}\right)^k\right]^{p/100}.$$

It is useful to know the behavior of the account if the number of times the interest is compounded per year becomes arbitrarily large (this is called *continuous compounding of interest*). Continuous compounding corresponds to calculating the limit of the last formula as k (or, equivalently, n), tends to infinity.

We know from the discussion in Subsection 6.2.3 that the expression $(1 + 1/k)^k$ tends to e. Therefore the size of the account after one year of continuous compounding of interest is

$$P \cdot e^{p/100}.$$

After t years of continuous compounding of interest the total money is

$$P \cdot e^{pt/100}. \tag{$\star\star$}$$

EXAMPLE 6.33

If \$6000 is placed in a savings account with 5% annual interest compounded continuously, then how large is the account after four and one half years?

SOLUTION

If $M(t)$ is the amount of money in the account at time t, then the preceding discussion guarantees that

$$M(t) = 6000 \cdot e^{5t/100}.$$

After four and one half years the size of the account is therefore

$$M(9/2) = 6000 \cdot e^{5 \cdot (9/2)/100} \approx \$7513.94.$$

EXAMPLE 6.34

A wealthy woman wishes to set up an endowment for her nephew. She wants the endowment to pay the young man \$100,000 in cash on the day of his twenty-first birthday. The endowment is set up on the day of the nephew's birth and is locked in at 11% interest compounded continuously. How much principal should be put into the account to yield the desired payoff?

SOLUTION

    Let P be the initial principal deposited in the account on the day of the nephew's birth. Using our compound interest equation ($\star\star$), we have

$$100000 = P \cdot e^{11 \cdot 21/100},$$

expressing the fact that after 21 years at 11% interest compounded continuously we want the value of the account to be $100,000.

    Solving for P gives

$$P = 100000 \cdot e^{-0.11 \cdot 21} = 100000 \cdot e^{-2.31} = 9926.13.$$

The aunt needs to endow the fund with an initial $9926.13.

You Try It: Suppose that we want a certain endowment to pay $50,000 in cash ten years from now. The endowment will be set up today with $5,000 principal and locked in at a fixed interest rate. What interest rate (compounded continuously) is needed to guarantee the desired payoff?

6.6   Inverse Trigonometric Functions

6.6.1   INTRODUCTORY REMARKS

Figure 6.14 shows the graphs of each of the six trigonometric functions. Notice that each graph has the property that some horizontal line intersects the graph at least twice. Therefore none of these functions is invertible. Another way of seeing this point is that each of the trigonometric functions is 2π-periodic (that is, the function repeats itself every 2π units: $f(x + 2\pi) = f(x)$), hence each of these functions is *not* one-to-one.

    If we want to discuss inverses for the trigonometric functions, then we must restrict their domains (this concept was introduced in Subsection 1.8.5). In this section we learn the standard methods for performing this restriction operation with the trigonometric functions.

6.6.2   INVERSE SINE AND COSINE

Consider the sine function with domain restricted to the interval $[-\pi/2, \pi/2]$ (Fig. 6.15). We use the notation $\mathrm{Sin}\, x$ to denote this restricted function. Observe that

$$\frac{d}{dx}\mathrm{Sin}\, x = \cos x > 0$$

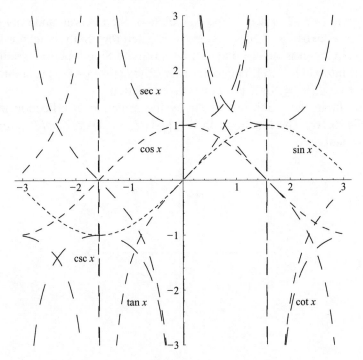

Fig. 6.14

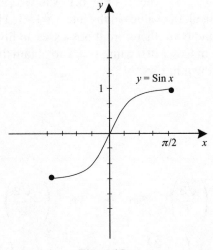

Fig. 6.15

on the interval $(-\pi/2, \pi/2)$. At the endpoints of the interval, and only there, the function $\text{Sin}\,x$ takes the values -1 and $+1$. Therefore $\text{Sin}\,x$ is increasing on its entire domain. So it is one-to-one. Furthermore the Sine function assumes every value in the interval $[-1, 1]$. Thus $\text{Sin} : [-\pi/2, \pi/2] \to [-1, 1]$ is one-to-one and onto; therefore $f(x) = \text{Sin}\,x$ is an invertible function.

We can obtain the graph of $\text{Sin}^{-1}x$ by the principle of reflection in the line $y = x$ (Fig. 6.16). The function $\text{Sin}^{-1} : [-1, 1] \to [-\pi/2, \pi/2]$ is increasing, one-to-one, and onto.

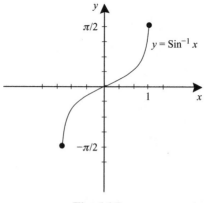

Fig. 6.16

The study of the inverse of cosine involves similar considerations, but we must select a different domain for our function. We define $\text{Cos}\,x$ to be the cosine function restricted to the interval $[0, \pi]$. Then, as Fig. 6.17 shows, $g(x) = \text{Cos}\,x$ is a one-to-one function. It takes on all the values in the interval $[-1, 1]$. Thus $\text{Cos} : [0, \pi] \to [-1, 1]$ is one-to-one and onto; therefore it possesses an inverse.

We reflect the graph of $\text{Cos}\,x$ in the line $y = x$ to obtain the graph of the function Cos^{-1}. The result is shown in Fig. 6.18.

EXAMPLE 6.35

Calculate

$$\text{Sin}^{-1}\left(\frac{\sqrt{3}}{2}\right), \quad \text{Sin}^{-1}0, \quad \text{Sin}^{-1}\left(-\frac{\sqrt{2}}{2}\right),$$

$$\text{Cos}^{-1}\left(-\frac{\sqrt{3}}{2}\right), \quad \text{Cos}^{-1}0, \quad \text{Cos}^{-1}\left(\frac{\sqrt{2}}{2}\right).$$

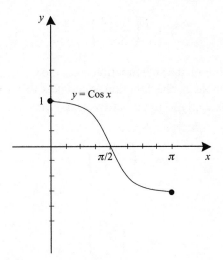

Fig. 6.17

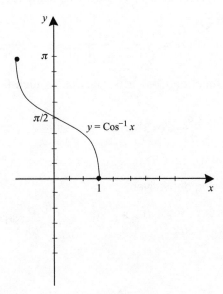

Fig. 6.18

SOLUTION
We have

$$\text{Sin}^{-1}\left(\frac{\sqrt{3}}{2}\right) = \frac{\pi}{3},$$

$$\text{Sin}^{-1} 0 = 0,$$

$$\text{Sin}^{-1}\left(-\frac{\sqrt{2}}{2}\right) = -\frac{\pi}{4}.$$

Notice that even though the sine function takes the value $\sqrt{3}/2$ at many different values of the variable x, the function Sine takes this value only at $x = \pi/3$. Similar comments apply to the other two examples.

We also have

$$\text{Cos}^{-1}\left(-\frac{\sqrt{3}}{2}\right) = \frac{5\pi}{6},$$

$$\text{Cos}^{-1} 0 = \frac{\pi}{2},$$

$$\text{Cos}^{-1}\left(\frac{\sqrt{2}}{2}\right) = \frac{\pi}{4}.$$

We calculate the derivative of $f(t) = \text{Sin}^{-1}t$ by using the usual trick for inverse functions. The result is

$$\frac{d}{dx}(\text{Sin}^{-1}(x)) = \frac{1}{\sqrt{1 - \sin^2(\text{Sin}^{-1}x)}} = \frac{1}{\sqrt{1 - x^2}}.$$

The derivative of the function $\text{Cos}^{-1}t$ is calculated much like that of $\text{Sin}^{-1}t$. We find that

$$\frac{d}{dx}(\text{Cos}^{-1}(x)) = -\frac{1}{\sqrt{1 - x^2}}.$$

EXAMPLE 6.36

Calculate the following derivatives:

$$\frac{d}{dx}\text{Sin}^{-1}x\bigg|_{x=\sqrt{2}/2}, \quad \frac{d}{dx}\text{Sin}^{-1}(x^2 + x)\bigg|_{x=1/3}, \quad \frac{d}{dx}\text{Sin}^{-1}\left(\frac{1}{x}\right)\bigg|_{x=-\sqrt{3}}.$$

SOLUTION

We have

$$\frac{d}{dx}\text{Sin}^{-1}x\bigg|_{x=\sqrt{2}/2} = \frac{1}{\sqrt{1 - x^2}}\bigg|_{x=\sqrt{2}/2} = \sqrt{2},$$

$$\frac{d}{dx}\text{Sin}^{-1}\left(x^2 + x\right)\bigg|_{x=1/3} = \frac{1}{\sqrt{1 - (x^2 + x)^2}} \cdot (2x + 1)\bigg|_{x=1/3} = \frac{15}{\sqrt{65}},$$

$$\frac{d}{dx}\text{Sin}^{-1}(1/x)\bigg|_{x=-\sqrt{3}} = \frac{1}{\sqrt{1 - (1/x)^2}} \cdot \left(-\frac{1}{x^2}\right)\bigg|_{x=-\sqrt{3}} = -\frac{1}{\sqrt{6}}.$$

You Try It: Calculate $(d/dx)\text{Cos}^{-1}[x^2 + x]$. Also calculate $(d/dx)\text{Sin}^{-1} \times [\ln x - x^3]$.

EXAMPLE 6.37

Calculate each of the following derivatives:

$$\frac{d}{dx}\text{Cos}^{-1} x\bigg|_{x=1/2}, \quad \frac{d}{dx}\text{Cos}^{-1}(\ln x)\bigg|_{x=\sqrt{e}}, \quad \frac{d}{dx}\text{Cos}^{-1}(\sqrt{x})\bigg|_{x=1/2}.$$

SOLUTION

We have

$$\frac{d}{dx}\text{Cos}^{-1}x\bigg|_{x=1/2} = -\frac{1}{\sqrt{1-x^2}}\bigg|_{x=1/2} = -\frac{2}{\sqrt{3}},$$

$$\frac{d}{dx}\text{Cos}^{-1}(\ln x)\bigg|_{x=\sqrt{e}} = -\frac{1}{\sqrt{1-(\ln x)^2}} \cdot \left(\frac{1}{x}\right)\bigg|_{x=\sqrt{e}} = -\frac{-2}{\sqrt{3e}},$$

$$\frac{d}{dx}\text{Cos}^{-1}(\sqrt{x})\bigg|_{x=1/2} = -\frac{1}{\sqrt{1-(\sqrt{x})^2}} \cdot \left(\frac{1}{2}x^{-1/2}\right)\bigg|_{x=1/2} = -1.$$

You Try It: Calculate $(d/dx)\ln[\text{Cos}^{-1}x]$ and $(d/dx)\exp[\text{Sin}^{-1}x]$.

6.6.3 THE INVERSE TANGENT FUNCTION

Define the function Tan x to be the restriction of tan x to the interval $(-\pi/2, \pi/2)$. Observe that the tangent function is undefined at the endpoints of this interval. Since

$$\frac{d}{dx}\text{Tan } x = \sec^2 x$$

we see that Tan x is increasing, hence it is one-to-one (Fig. 6.19). Also Tan takes arbitrarily large positive values when x is near to, but less than, $\pi/2$. And Tan takes negative values that are arbitrarily large in absolute value when x is near to, but greater than, $-\pi/2$. Therefore Tan takes all real values. Since Tan : $(-\pi/2, \pi/2) \to (-\infty, \infty)$ is one-to-one and onto, the inverse function Tan^{-1} : $(-\infty, \infty) \to (-\pi/2, \pi/2)$ exists. The graph of this inverse function is shown in Fig. 6.20. It is obtained by the usual procedure of reflecting in the line $y = x$.

EXAMPLE 6.38

Calculate

$$\text{Tan}^{-1} 1, \quad \text{Tan}^{-1} 1/\sqrt{3}, \quad \text{Tan}^{-1}(-\sqrt{3}).$$

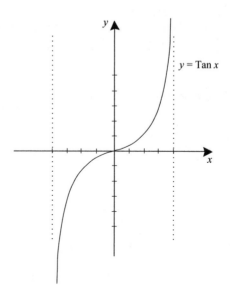

Fig. 6.19

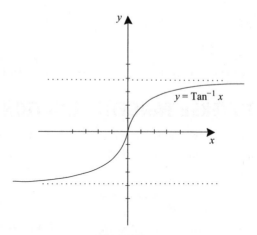

Fig. 6.20

SOLUTION
We have

$$\text{Tan}^{-1}1 = \frac{\pi}{4},$$
$$\text{Tan}^{-1}1/\sqrt{3} = \frac{\pi}{6},$$
$$\text{Tan}^{-1}(-\sqrt{3}) = -\frac{\pi}{3}.$$

As with the first two trigonometric functions, we note that the tangent function takes each of the values $1, 1/\sqrt{3}, -\sqrt{3}$ at many different points of its domain. But Tan x takes each of these values at just one point of its domain.

The derivative of our new function may be calculated in the usual way. The result is

$$\frac{d}{dt}\text{Tan}^{-1}t = \frac{1}{1+t^2}.$$

Next we calculate some derivatives:

EXAMPLE 6.39

Calculate the following derivatives:

$$\frac{d}{dx}\text{Tan}^{-1}x\bigg|_{x=1}, \quad \frac{d}{dx}\text{Tan}^{-1}(x^3)\bigg|_{x=\sqrt{2}}, \quad \frac{d}{dx}\text{Tan}^{-1}(e^x)\bigg|_{x=0}.$$

SOLUTION
We have

$$\frac{d}{dx}\text{Tan}^{-1}x\bigg|_{x=1} = \frac{1}{1+x^2}\bigg|_{x=1} = \frac{1}{2},$$

$$\frac{d}{dx}\text{Tan}^{-1}(x^3)\bigg|_{x=\sqrt{2}} = \frac{1}{1+(x^3)^2}\cdot 3x^2\bigg|_{x=\sqrt{2}} = \frac{2}{3},$$

$$\frac{d}{dx}\text{Tan}^{-1}(e^x)\bigg|_{x=0} = \frac{1}{1+(e^x)^2}\cdot e^x\bigg|_{x=0} = \frac{1}{2}.$$

You Try It: Calculate $(d/dx)\text{Tan}^{-1}[\ln x + x^3]$ and $(d/dx)\ln[\text{Tan}^{-1}x]$.

6.6.4   INTEGRALS IN WHICH INVERSE TRIGONOMETRIC FUNCTIONS ARISE

Our differentiation formulas for inverse trigonometric functions can be written in reverse, as antidifferentiation formulas. We have

$$\int \frac{du}{\sqrt{1-u^2}} = \text{Sin}^{-1}u + C;$$

$$\int \frac{du}{\sqrt{1-u^2}} = -\text{Cos}^{-1}u + C;$$

$$\int \frac{du}{1+u^2}du = \text{Tan}^{-1}u + C.$$

The important lesson here is that, while the integrands involve only polynomials and roots, the antiderivatives involve inverse trigonometric functions.

EXAMPLE 6.40

Evaluate the integral

$$\int \frac{\sin x}{1 + \cos^2 x} \, dx.$$

SOLUTION

For clarity we set $\varphi(x) = \cos x$, $\varphi'(x) = -\sin x$. The integral becomes

$$-\int \frac{\varphi'(x) \, dx}{1 + \varphi^2(x)}.$$

By what we have just learned about Tan^{-1}, this last integral is equal to

$$-\mathrm{Tan}^{-1}\varphi(x) + C.$$

Resubstituting $\varphi(x) = \cos x$ yields that

$$\int \frac{\sin x}{1 + \cos^2 x} \, dx = -\mathrm{Tan}^{-1}(\cos x) + C.$$

You Try It: Calculate $\int x/(1 + x^4) \, dx$.

EXAMPLE 6.41

Calculate the integral

$$\int \frac{3x^2}{\sqrt{1 - x^6}} \, dx.$$

SOLUTION

For clarity we set $\varphi(x) = x^3$, $\varphi'(x) = 3x^2$. The integral then becomes

$$\int \frac{\varphi'(x) \, dx}{\sqrt{1 - \varphi^2(x)}}.$$

We know that this last integral equals

$$\mathrm{Sin}^{-1}\varphi(x) + C.$$

Resubstituting the formula for φ gives a final answer of

$$\int \frac{3x^2}{\sqrt{1 - x^6}} \, dx = \mathrm{Sin}^{-1}(x^3) + C.$$

You Try It: Evaluate the integral

$$\int \frac{x \, dx}{\sqrt{1 - x^4}}.$$

6.6.5   OTHER INVERSE TRIGONOMETRIC FUNCTIONS

The most important inverse trigonometric functions are Sin^{-1}, Cos^{-1}, and Tan^{-1}. We say just a few words about the other three.

Define $\mathrm{Cot}\,x$ to be the restriction of the cotangent function to the interval $(0, \pi)$ (Fig. 6.21). Then Cot is decreasing on that interval and takes on all real values. Therefore the inverse

$$\mathrm{Cot}^{-1} : (-\infty, \infty) \to (0, \pi)$$

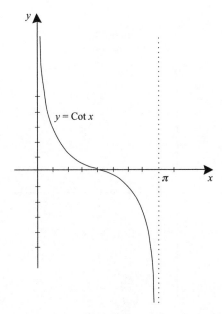

$y = \mathrm{Cot}\,x$

Fig. 6.21

is well defined. Look at Fig. 6.22 for the graph. It can be shown that

$$\frac{d}{dx}\mathrm{Cot}^{-1}\,x = -\frac{1}{1+x^2}.$$

Define $\mathrm{Sec}\,x$ to be the function $\sec x$ restricted to the set $[0, \pi/2) \cup (\pi/2, \pi]$ (Fig. 6.23). Then $\mathrm{Sec}\,x$ is one-to-one. For these values of the variable x, the cosine function takes all values in the interval $[-1, 1]$ except for 0. Passing to the reciprocal, we see that secant takes all values greater than or equal to 1 and all values less than or equal to -1. The inverse function is

$$\mathrm{Sec}^{-1} : (-\infty, -1] \cup [1, \infty) \to [0, \pi/2) \cup (\pi/2, \pi]$$

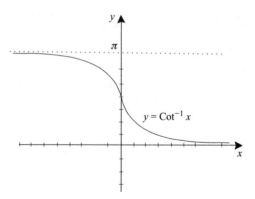

Fig. 6.22

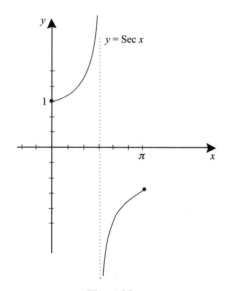

Fig. 6.23

(Fig. 6.24). It can be shown that

$$\frac{d}{dx}\operatorname{Sec}^{-1}x = \frac{1}{|x| \cdot \sqrt{x^2 - 1}}, \quad |x| > 1.$$

The function Csc x is defined to be the restriction of Csc x to the set $[-\pi/2, 0) \cup (0, \pi/2]$. The graph is exhibited in Fig. 6.25. Then Csc x is one-to-one. For these values of the x variable, the sine function takes on all values in the interval $[-1, 1]$ except for 0. Therefore Csc takes on all values greater than or equal to 1 and all values less than or equal to -1; Csc^{-1} therefore has domain $(-\infty, -1] \cup [1, \infty)$ and takes values in $[-1, 0) \cup (0, 1]$ (Fig. 6.26).

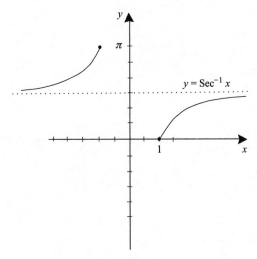

Fig. 6.24

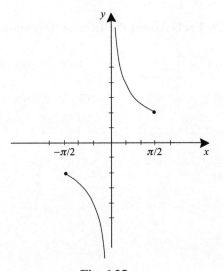

Fig. 6.25

It is possible to show that

$$\frac{d}{dx}\mathrm{Csc}^{-1}x = -\frac{1}{|x| \cdot \sqrt{x^2 - 1}}, \quad |x| > 1.$$

You Try It: What is $\mathrm{Sec}^{-1}(-2/\sqrt{3})$? What is $\mathrm{Csc}^{-1}(-\sqrt{2})$?

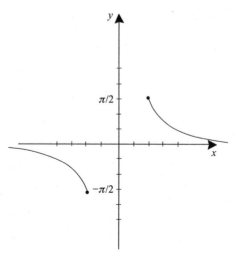

Fig. 6.26

Summary of Key Facts About the Inverse Trigonometric Functions

$$\text{Sin}\, x = \sin x, -\frac{\pi}{2} \le x \le \frac{\pi}{2}; \quad \text{Cos}\, x = \cos x, 0 \le x \le \pi;$$

$$\text{Tan}\, x = \tan x, -\frac{\pi}{2} < x < \frac{\pi}{2}; \quad \text{Cot}\, x = \cot x, 0 < x < \pi;$$

$$\text{Sec}\, x = \sec x, x \in [0, \pi/2) \cup (\pi/2, \pi]; \quad \text{Csc}\, x = \csc x, x \in [-\pi/2, 0) \cup (0, \pi/2].$$

$$\frac{d}{dx}\text{Sin}^{-1}x = \frac{1}{\sqrt{1-x^2}}, -1 < x < 1; \quad \frac{d}{dx}\text{Cos}^{-1}x = -\frac{1}{\sqrt{1-x^2}}, -1 < x < 1;$$

$$\frac{d}{dx}\text{Tan}^{-1}x = \frac{1}{1+x^2}, -\infty < x < \infty; \quad \frac{d}{dx}\text{Cot}^{-1}x = -\frac{1}{1+x^2}, -\infty < x < \infty;$$

$$\frac{d}{dx}\text{Sec}^{-1}x = \frac{1}{|x| \cdot \sqrt{x^2-1}}, |x| > 1; \quad \frac{d}{dx}\text{Csc}^{-1}x = -\frac{1}{|x| \cdot \sqrt{x^2-1}}, |x| > 1;$$

$$\int \frac{du}{\sqrt{1-u^2}} = \text{Sin}^{-1}u + C; \quad \int \frac{du}{\sqrt{1-u^2}} = -\text{Cos}^{-1}u + C;$$

$$\int \frac{du}{1+u^2}du = \text{Tan}^{-1}u + C; \quad \int \frac{du}{1+u^2}du = -\text{Cot}^{-1}u + C;$$

$$\int \frac{du}{|u| \cdot \sqrt{u^2-1}} = \text{Sec}^{-1}u + C; \quad \int \frac{du}{|u| \cdot \sqrt{u^2-1}} = -\text{Csc}^{-1}u + C.$$

You Try It: What is the derivative of $\text{Sec}^{-1}x^2$?

6.6.6   AN EXAMPLE INVOLVING INVERSE TRIGONOMETRIC FUNCTIONS

EXAMPLE 6.42

Hypatia is viewing a ten-foot-long tapestry that is hung lengthwise on a wall. The bottom end of the tapestry is two feet above her eye level. At what distance should she stand from the tapestry in order to obtain the most favorable view?

SOLUTION

For the purposes of this problem, view A is considered more favorable than view B if it provides a greater sweep for the eyes. In other words, form the triangle with vertices (i) the eye of the viewer, (ii) the top of the tapestry, and (iii) the bottom of the tapestry (Fig. 6.27). Angle α is the angle at the eye of the viewer. We want the viewer to choose her position so that the angle α at the eye of the viewer is maximized.

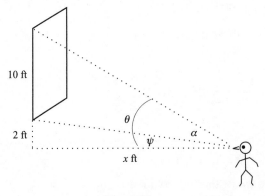

Fig. 6.27

The figure shows a mathematical model for the problem. The angle α is the angle θ less the angle ψ. Thus we have

$$\alpha = \theta - \psi = \mathrm{Cot}^{-1}(x/12) - \mathrm{Cot}^{-1}(x/2).$$

Notice that when the viewer is standing with her face against the wall then $\theta = \psi = \pi/2$ so that $\alpha = 0$. Also when the viewer is far from the tapestry then $\theta - \alpha$ is quite small. So the maximum value for α will occur for some finite, positive value of x. That value can be found by differentiating α with respect to x, setting the derivative equal to zero, and solving for x.

We leave it to you to perform the calculation and discover that $\sqrt{24}$ ft is the optimal distance at which the viewer should stand.

You Try It: Redo the last example if the tapestry is 20 feet high and the bottom of the tapestry is 6 inches above eye level.

Exercises

1. Simplify these logarithmic expressions.

    (a) $\ln \dfrac{a^2 \cdot b^{-3}}{c^4 \cdot d}$

    (b) $\dfrac{\log_3(a^2 b)}{\log_2(a^2 b)}$

    (c) $\ln[e^{3x} \cdot z^4 \cdot w^{-3}]$

    (d) $\log_{10}[100^w \cdot \sqrt{10}]$

2. Solve each of these equations for x.

    (a) $3^x \cdot 5^{-x} = 2^x \cdot e^3$

    (b) $\dfrac{3^x}{5^{-x} \cdot 4^{2x}} = 10^x \cdot 10^2$

    (c) $2^{3x} \cdot 3^{4x} \cdot 4^{5x} = 6$

    (d) $\dfrac{5}{2^{3x} \cdot e^{5x}} = \dfrac{2}{2^x \cdot 3^{-x}}$

3. Calculate each of these derivatives.

    (a) $\dfrac{d}{dx} \ln[\sin(x^2)]$

    (b) $\dfrac{d}{dx} \ln\left[\dfrac{x^2}{x-1}\right]$

    (c) $\dfrac{d}{dx} e^{\sin(e^x)}$

    (d) $\dfrac{d}{dx} \sin(\ln x)$

4. Calculate each of these integrals.

    (a) $\displaystyle\int e^{-x} x^3 \, dx$ [*Hint*: Guess $p(x) \cdot e^{-x}$, p a polynomial.]

(b) $\displaystyle\int x^2 \cdot \ln^2 x \, dx$ [*Hint*: Guess $p(x) \cdot \ln^2 x + q(x) \ln x + r(x)$, p, q, r polynomials.]

(c) $\displaystyle\int_1^e \frac{\ln x}{x} \, dx$

(d) $\displaystyle\int_1^2 \frac{e^x}{e^x + 1} \, dx$

5. Use the technique of logarithmic differentiation to calculate the derivative of each of the following functions.

(a) $x^3 \cdot \dfrac{x^2 + 1}{x^3 - x}$

(b) $\dfrac{\sin x \cdot (x^3 + x)}{x^2(x + 1)}$

(c) $(x^2 + x^3)^4 \cdot (x^2 + x)^{-3} \cdot (x - 1)$

(d) $\dfrac{x \cdot \cos x}{\ln x \cdot e^x}$

6. There are 5 grams of a certain radioactive substance present at noon on January 10 and 3 grams present at noon on February 10. How much will be present at noon on March 10?

7. A petri dish has 10,000 bacteria present at 10:00 a.m. and 15,000 present at 1:00 p.m. How many bacteria will there be at 2:00 p.m.?

8. A sum of \$1000 is deposited on January 1, 2005 at 6% annual interest, compounded continuously. All interest is re-invested. How much money will be in the account on January 1, 2009?

9. Calculate these derivatives.

(a) $\dfrac{d}{dx} \mathrm{Sin}^{-1}(x \cdot e^x)$

(b) $\dfrac{d}{dx} \mathrm{Tan}^{-1}\left(\dfrac{x}{x + 1}\right)$

(c) $\dfrac{d}{dx} \mathrm{Tan}^{-1}[\ln(x^2 + x)]$

(d) $\dfrac{d}{dx} \mathrm{Sec}^{-1}(\tan x)$

10. Calculate each of these integrals.

(a) $\displaystyle\int \frac{2x}{1 + x^4} x \, dx$

(b) $\displaystyle\int \frac{3x^2}{\sqrt{1-x^6}}\,dx$

(c) $\displaystyle\int_0^{\pi/2} \frac{2\sin x \cos x}{\sqrt{1-\sin^4 x}}\,dx$

(d) $\displaystyle\int \frac{dx}{5+2x^2}$

CHAPTER 7

Methods of Integration

7.1 Integration by Parts

We learned in Section 4.5 that the integral of the sum of two functions is the sum of the respective integrals. But what of the integral of a product? The following reasoning is incorrect:

$$\int x^2 \, dx = \int x \cdot x \, dx = \int x \, dx \cdot \int x \, dx$$

because the left-hand side is $x^3/3$ while the right-hand side is $(x^2/2) \cdot (x^2/2) = x^4/4$.

The correct technique for handling the integral of a product is a bit more subtle, and is called *integration by parts*. It is based on the product rule

$$(u \cdot v)' = u' \cdot v + u \cdot v'.$$

Integrating both sides of this equation, we have

$$\int (u \cdot v)' \, dx = \int u' \cdot v \, dx + \int u \cdot v' \, dx.$$

The Fundamental Theorem of Calculus tells us that the left-hand side is $u \cdot v$. Thus

$$u \cdot v = \int u' \cdot v \, dx + \int u \cdot v' \, dx$$

or

$$\int u \cdot v' \, dx = u \cdot v - \int v \cdot u' \, dx.$$

It is traditional to abbreviate $u'(x) \, dx = du$ and $v'(x) \, dx = dv$. Thus the integration by parts formula becomes

$$\int u \, dv = uv - \int v \, du.$$

Let us now learn how to use this simple new formula.

EXAMPLE 7.1

Calculate

$$\int x \cdot \cos x \, dx.$$

SOLUTION

We observe that the integrand is a product. Let us use the integration by parts formula by setting $u(x) = x$ and $dv = \cos x \, dx$. Then

$$u(x) = x \qquad du = u'(x) \, dx = 1 \, dx = dx$$
$$v(x) = \sin x \quad dv = v'(x) \, dx = \cos x \, dx$$

Of course we calculate v by anti-differentiation.

According to the integration by parts formula,

$$\int x \cdot \cos x \, dx = \int u \, dv$$
$$= u \cdot v - \int v \, du$$
$$= x \cdot \sin x - \int \sin x \, dx$$
$$= x \cdot \sin x - (-\cos x) + C$$
$$= x \cdot \sin x + \cos x + C.$$

Math Note: Observe that we can check the answer in the last example just by differentiation:

$$\frac{d}{dx}[x \cdot \sin x + \cos x + C] = 1 \cdot \sin x + x \cdot \cos x - \sin x = x \cdot \cos x.$$

The choice of u and v in the integration by parts technique is significant. We selected u to be x because then du will be $1 \, dx$, thereby simplifying the integral.

If we had instead selected $u = \cos x$ and $dv = x\, dx$ then we would have found
that $v = x^2/2$ and $du = -\sin x\, dx$ and the new integral

$$\int v\, du = \int \frac{x^2}{2}(-\sin x)\, dx$$

is more complicated.

EXAMPLE 7.2

Calculate the integral

$$\int x^2 \cdot e^x\, dx.$$

SOLUTION

Keeping in mind that we want to choose u and v so as to simplify the integral,
we take $u = x^2$ and $dv = e^x\, dx$. Then

$$u(x) = x^2 \quad du = u'(x)\, dx = 2x\, dx$$
$$v(x) = e^x \quad dv = v'(x)\, dx = e^x\, dx$$

Then the integration by parts formula tells us that

$$\int x^2 e^x\, dx = \int u\, dv = uv - \int v\, du = x^2 \cdot e^x - \int e^x \cdot 2x\, dx. \qquad (*)$$

We see that we have transformed the integral into a simpler one (involving
$x \cdot e^x$ instead of $x^2 \cdot e^x$), but another integration by parts will be required. Now
we take $u = 2x$ and $dv = e^x\, dx$. Then

$$u(x) = 2x \quad du = u'(x)\, dx = 2\, dx$$
$$v(x) = e^x \quad dv = v'(x)\, dx = e^x\, dx$$

So equation $(*)$ equals

$$x^2 \cdot e^x - \int u\, dv = x^2 \cdot e^x - \left[u \cdot v - \int v\, du \right]$$
$$= x^2 \cdot e^x - \left[2x \cdot e^x - \int e^x \cdot 2\, dx \right]$$
$$= x^2 \cdot e^x - 2x \cdot e^x + 2e^x + C.$$

We leave it to the reader to check this last answer by differentiation.

You Try It: Calculate the integral

$$\int x^2 \log x\, dx.$$

EXAMPLE 7.3

Calculate

$$\int_1^2 \log x \, dx.$$

SOLUTION

This example differs from the previous ones because now we are evaluating a *definite integral* (i.e., an integral with numerical limits). We still use the integration by parts formula, keeping track of the numerical limits of integration.

We first notice that, on the one hand, the integrand is *not* a product. On the other hand, we certainly do not know an antiderivative for $\log x$. We remedy the situation by writing $\log x = 1 \cdot \log x$. Now the only reasonable choice is to take $u = \log x$ and $dv = 1 \, dx$. Therefore

$$u(x) = \log x \quad du = u'(x) \, dx = (1/x) \, dx$$
$$v(x) = x \quad dv = v'(x) \, dx = 1 \, dx$$

and

$$\int_1^2 1 \cdot \log x \, dx = \int_1^2 u \, dv$$

$$= uv \Big|_1^2 - \int_1^2 v \, du$$

$$= (\log x) \cdot x \Big|_1^2 - \int_1^2 x \cdot \frac{1}{x} \, dx$$

$$= 2 \cdot \log 2 - 1 \cdot \log 1 - \int_1^2 1 \, dx$$

$$= 2 \cdot \log 2 - x \Big|_1^2$$

$$= 2 \cdot \log 2 - (2 - 1)$$

$$= 2 \cdot \log 2 - 1.$$

You Try It: Evaluate

$$\int_0^4 x^2 \cdot \sin x \, dx.$$

We conclude this section by doing another definite integral, but we use a slightly different approach from that in Example 7.3.

EXAMPLE 7.4

Calculate the integral

$$\int_{\pi/2}^{2\pi} \sin x \cos x \, dx.$$

SOLUTION

We use integration by parts, but we apply the technique to the corresponding indefinite integral. We let $u = \sin x$ and $dv = \cos x \, dx$. Then

$$u(x) = \sin x \quad du = u'(x) \, dx = \cos x \, dx$$
$$v(x) = \sin x \quad dv = v'(x) \, dx = \cos x \, dx$$

So

$$\int \sin x \cos x \, dx = \int u \, dv$$

$$= uv - \int v \, du$$

$$= (\sin x) \cdot (\sin x) - \int \sin x \cos x \, dx.$$

At first blush, it appears that we have accomplished nothing. For the new integral is just the same as the old integral. But in fact we can move the new integral (on the right) to the left-hand side to obtain

$$2 \int \sin x \cos x \, dx = \sin^2 x.$$

Throwing in the usual constant of integration, we obtain

$$\int \sin x \cos x \, dx = \frac{1}{2} \sin^2 x + C.$$

Now we complete our work by evaluating the definite integral:

$$\int_{\pi/2}^{2\pi} \sin x \cos x \, dx = \frac{1}{2} \sin^2 x \Big|_{\pi/2}^{2\pi} = \frac{1}{2} \left[\sin^2 2\pi - \sin^2 (\pi/2) \right] = -\frac{1}{2}.$$

We see that there are two ways to treat a definite integral using integration by parts. One is to carry the limits of integration along with the parts calculation. The other is to do the parts calculation first (with an indefinite integral) and then plug in the limits of integration at the end. Either method will lead to the same solution.

You Try It: Calculate the integral

$$\int_{0}^{2} e^{-x} \cos 2x \, dx.$$

7.2 Partial Fractions

7.2.1 INTRODUCTORY REMARKS

The method of partial fractions is used to integrate rational functions, or quotients of polynomials. We shall treat here some of the basic aspects of the technique.

The first fundamental observation is that there are some elementary rational functions whose integrals we already know.

I   Integrals of Reciprocals of Linear Functions   An integral

$$\int \frac{1}{ax + b}\, dx$$

with $a \neq 0$ is always a logarithmic function. In fact we can calculate

$$\int \frac{1}{ax + b}\, dx = \frac{1}{a} \int \frac{1}{x + b/a}\, dx = \frac{1}{a} \log|x + b/a|.$$

II   Integrals of Reciprocals of Quadratic Expressions   An integral

$$\int \frac{1}{c + ax^2}\, dx,$$

when a and c are positive, is an inverse trigonometric function. In fact we can use what we learned in Section 6.6.3 to write

$$\int \frac{1}{c + ax^2}\, dx = \frac{1}{c} \int \frac{1}{1 + (a/c)x^2}\, dx$$

$$= \frac{1}{c} \int \frac{1}{1 + (\sqrt{a/c}\,x)^2}\, dx$$

$$= \frac{1}{\sqrt{ac}} \frac{\sqrt{a}}{\sqrt{c}} \int \frac{1}{1 + (\sqrt{a/c}\,x)^2}\, dx$$

$$= \frac{1}{\sqrt{ac}} \arctan\left(\sqrt{a/c}\,x\right) + C.$$

III   More Integrals of Reciprocals of Quadratic Expressions   An integral

$$\int \frac{1}{ax^2 + bx + c}\, dx$$

with $a > 0$, and discriminant $b^2 - 4ac$ negative, will also be an inverse trigonometric function. To see this, we notice that we can write

$$ax^2 + bx + c = a\left(x^2 + \frac{b}{a}x + \quad\right) + c$$

$$= a\left(x^2 + \frac{b}{a}x + \frac{b^2}{4a^2}\right) + \left(c - \frac{b^2}{4a}\right)$$

$$= a \cdot \left(x + \frac{b}{2a}\right)^2 + \left(c - \frac{b^2}{4a}\right).$$

Since $b^2 - 4ac < 0$, the final expression in parentheses is positive. For simplicity, let $\lambda = b/2a$ and let $\gamma = c - b^2/(4a)$. Then our integral is

$$\int \frac{1}{\gamma + a \cdot (x + \lambda)^2}\, dx.$$

Of course we can handle this using **II** above. We find that

$$\int \frac{1}{ax^2 + bx + c}\, dx = \int \frac{1}{\gamma + a \cdot (x + \lambda)^2}\, dx$$

$$= \frac{1}{\sqrt{a\gamma}} \cdot \arctan\left(\frac{\sqrt{a}}{\sqrt{\gamma}} \cdot (x + \lambda)\right) + C.$$

IV   Even More on Integrals of Reciprocals of Quadratic Expressions
Evaluation of the integral

$$\int \frac{1}{ax^2 + bx + c}\, dx$$

when the discriminant $b^2 - 4ac$ is ≥ 0 will be a consequence of the work we do below with partial fractions. We will say no more about it here.

7.2.2   PRODUCTS OF LINEAR FACTORS

We illustrate the technique of partial fractions by way of examples.

EXAMPLE 7.5

Here we treat the case of distinct linear factors.
   Let us calculate

$$\int \frac{1}{x^2 - 3x + 2}\, dx.$$

SOLUTION

We notice that the integrand factors as

$$\frac{1}{x^2 - 3x + 2} = \frac{1}{(x-1)(x-2)}. \qquad (*)$$

[Notice that the quadratic polynomial in the denominator will factor *precisely* *when* the discriminant is ≥ 0, which is case **IV** from Subsection 7.2.1.] Our goal is to write the fraction on the right-hand side of $(*)$ as a sum of simpler fractions. With this thought in mind, we write

$$\frac{1}{(x-1)(x-2)} = \frac{A}{x-1} + \frac{B}{x-2},$$

where A and B are constants to be determined. Let us put together the two fractions on the right by placing them over the common denominator $(x-1) \times (x-2)$. Thus

$$\frac{1}{(x-1)(x-2)} = \frac{A}{x-1} + \frac{B}{x-2} = \frac{A(x-2) + B(x-1)}{(x-1)(x-2)}.$$

The only way that the fraction on the far left can equal the fraction on the far right is if their numerators are equal. This observation leads to the equation

$$1 = A(x-2) + B(x-1)$$

or

$$0 = (A+B)x + (-2A - B - 1).$$

Now this equation is to be identically true in x; in other words, it must hold for every value of x. So the coefficients must be 0.

At long last, then, we have a system of two equations in two unknowns:

$$A + B = 0$$
$$-2A - B - 1 = 0$$

Of course this system is easily solved and the solutions found to be $A = -1$, $B = 1$. We conclude that

$$\frac{1}{(x-1)(x-2)} = \frac{-1}{x-1} + \frac{1}{x-2}.$$

What we have learned, then, is that

$$\int \frac{1}{x^2 - 3x + 2} \, dx = \int \frac{-1}{x-1} \, dx + \int \frac{1}{x-2} \, dx.$$

Each of the individual integrals on the right may be evaluated using the information in **I** of Subsection 7.2.1. As a result,

$$\int \frac{1}{x^2 - 3x + 2} \, dx = -\log |x - 1| + \log |x - 2| + C.$$

You Try It: Calculate the integral

$$\int_1^4 \frac{dx}{x^2 + 5x + 4}.$$

Now we consider repeated linear factors.

EXAMPLE 7.6

Let us evaluate the integral

$$\int \frac{dx}{x^3 - 4x^2 - 3x + 18}.$$

SOLUTION

In order to apply the method of partial fractions, we first must factor the denominator of the integrand. It is known that every polynomial with real coefficients will factor into linear and quadratic factors. How do we find this factorization? Of course we must find a root. For a polynomial of the form

$$x^k + a_{k-1}x^{k-1} + a_{k-2}x^{k-2} + \cdots + a_1 x + a_0,$$

any integer root will be a factor of a_0. This leads us to try $\pm 1, \pm 2, \pm 3, \pm 6, \pm 9$ and ± 18. We find that -2 and 3 are roots of $x^3 - 4x^2 - 3x + 18$. In point of fact,

$$x^3 - 4x^2 - 3x + 18 = (x + 2) \cdot (x - 3)^2.$$

An attempt to write

$$\frac{1}{x^3 - 4x^2 - 3x + 18} = \frac{A}{x + 2} + \frac{B}{x - 3}$$

will not work. We encourage the reader to try this for himself so that he will understand why an extra idea is needed.

In fact we will use the paradigm

$$\frac{1}{x^3 - 4x^2 - 3x + 18} = \frac{A}{x + 2} + \frac{B}{x - 3} + \frac{C}{(x - 3)^2}.$$

Putting the right-hand side over a common denominator yields

$$\frac{1}{x^3 - 4x^2 - 3x + 18} = \frac{A(x - 3)^2 + B(x + 2)(x - 3) + C(x + 2)}{x^3 - 4x^2 - 3x + 18}.$$

Of course the numerators must be equal, so

$$1 = A(x-3)^2 + B(x+2)(x-3) + C(x+2).$$

We rearrange the equation as

$$(A+B)x^2 + (-6A - B + C)x + (9A - 6B + 2C - 1) = 0.$$

Since this must be an identity in x, we arrive at the system of equations

$$
\begin{aligned}
A + B &= 0 \\
-6A - B + C &= 0 \\
9A - 6B + 2C - 1 &= 0
\end{aligned}
$$

This system is easily solved to yield $A = 1/25$, $B = -1/25$, $C = 1/5$.

As a result of these calculations, our integral can be transformed as follows:

$$\int \frac{1}{x^3 - 4x^2 - 3x + 18}\, dx = \int \frac{1/25}{x+2}\, dx + \int \frac{-1/25}{x-3}\, dx + \int \frac{1/5}{(x-3)^2}\, dx.$$

The first integral equals $(1/25) \log |x+2|$, the second integral equals $-(1/25) \log |x-3|$, and the third integral equals $-(1/5)/(x-3)$.

In summary, we have found that

$$\int \frac{1}{x^3 - 4x^2 - 3x + 18}\, dx = \frac{\log|x+2|}{25} - \frac{\log|x-3|}{25} - \frac{1}{5(x-3)} + C.$$

You Try It: Evaluate the integral

$$\int_2^4 \frac{x\, dx}{x^3 + 5x^2 + 7x + 3}.$$

7.2.3   QUADRATIC FACTORS

EXAMPLE 7.7

Evaluate the integral

$$\int \frac{x\, dx}{x^3 + 2x^2 + x + 2}.$$

SOLUTION

Since the denominator is a cubic polynomial, it must factor. The factors of the constant term are ± 1 and ± 2. After some experimentation, we find that $x = -2$ is a root and in fact the polynomial factors as

$$x^3 + 2x^2 + x + 2 = (x+2)(x^2+1).$$

Thus we wish to write the integrand as the sum of a factor with denominator $(x + 2)$ and another factor with denominator $(x^2 + 1)$. The correct way to do this is

$$\frac{x}{x^3 + 2x^2 + x + 2} = \frac{x}{(x + 2)(x^2 + 1)} = \frac{A}{x + 2} + \frac{Bx + C}{x^2 + 1}.$$

We put the right-hand side over a common denominator to obtain

$$\frac{x}{x^3 + 2x^2 + x + 2} = \frac{A(x^2 + 1) + (Bx + C)(x + 2)}{x^3 + 2x^2 + x + 2}.$$

Identifying numerators leads to

$$x = (A + B)x^2 + (2B + C)x + (A + 2C).$$

This equation must be identically true, so we find (identifying powers of x) that

$$A + \quad B \qquad = 0$$
$$2B + \quad C = 1$$
$$A \qquad + \quad 2C = 0$$

Solving this system, we find that $A = -2/5$, $B = 2/5$, $C = 1/5$. So

$$\int \frac{x\,dx}{x^3 + 2x^2 + x + 2} = \int \frac{-2/5}{x + 2}\,dx + \int \frac{(2/5)x + (1/5)}{x^2 + 1}\,dx$$
$$= \frac{-2}{5}\log|x + 2| + \frac{1}{5}\int \frac{2x}{x^2 + 1}\,dx + \frac{1}{5}\int \frac{dx}{x^2 + 1}\,dx$$
$$= \frac{-2}{5}\log|x + 2| + \frac{1}{5}\log|x^2 + 1| + \frac{1}{5}\arctan x + C.$$

You Try It: Calculate the integral

$$\int_0^1 \frac{dx}{x^3 + 6x^2 + 9x}.$$

You Try It: Calculate the integral

$$\int \frac{dx}{x^3 + x}.$$

7.3   Substitution

Sometimes it is convenient to transform a given integral into another one by means of a change of variable. This method is often called "the method of change of variable" or "u-substitution."

To see a model situation, imagine an integral

$$\int_a^b f(x)\, dx.$$

If the techniques that we know will not suffice to evaluate the integral, then we might attempt to transform this to another integral by a change of variable $x = \varphi(t)$. This entails $dx = \varphi'(t)\, dt$. Also

$$x = a \longleftrightarrow t = \varphi^{-1}(a) \qquad \text{and} \qquad x = b \longleftrightarrow t = \varphi^{-1}(b).$$

Thus the original integral is transformed to

$$\int_{\varphi^{-1}(a)}^{\varphi^{-1}(b)} f(\varphi(t)) \cdot \varphi'(t)\, dt.$$

It turns out that, with a little notation, we can make this process both convenient and straightforward.

We now illustrate this new paradigm with some examples. We begin with an indefinite integral.

EXAMPLE 7.8

Evaluate

$$\int [\sin x]^5 \cdot \cos x\, dx.$$

SOLUTION

On looking at the integral, we see that the expression $\cos x$ is the derivative of $\sin x$. This observation suggests the substitution $\sin x = u$. Thus $\cos x\, dx = du$. We must now substitute these expressions into the integral, replacing all x-expressions with u-expressions. When we are through with this process, no x expressions can remain. The result is

$$\int u^5\, du.$$

This is of course an easy integral for us. So we have

$$\int [\sin x]^5 \cdot \cos x\, dx = \int u^5\, du = \frac{u^6}{6} + C.$$

Now the important final step is to resubstitute the x-expressions in place of the u-expressions. The result is then

$$\int [\sin x]^5 \cdot \cos x\, dx = \frac{\sin^6 x}{6} + C.$$

Math Note: Always be sure to check your work. You can differentiate the answer in the last example to recover the integrand, confirming that the integration has been performed correctly.

EXAMPLE 7.9

Evaluate the integral

$$\int_0^3 2x\sqrt{x^2 + 1}\ dx.$$

SOLUTION

We recognize that the expression $2x$ is the derivative of $x^2 + 1$. This suggests the substitution $u = x^2 + 1$. Thus $du = 2x\ dx$. Also $x = 0 \longleftrightarrow u = 1$ and $x = 3 \longleftrightarrow u = 10$. The integral is thus transformed to

$$\int_1^{10} \sqrt{u}\ du.$$

This new integral is a bit easier to understand if we write the square root as a fractional power:

$$\int_1^{10} u^{1/2}\ du = \frac{u^{3/2}}{3/2}\Big|_1^{10} = \frac{10^{3/2}}{3/2} - \frac{1^{3/2}}{3/2} = \frac{2 \cdot 10^{3/2}}{3} - \frac{2}{3}.$$

You Try It: Evaluate the integral

$$\int_3^5 \frac{dx}{x \cdot \log |x|}.$$

Math Note: Just as with integration by parts, we always have the option of first evaluating the indefinite integral and then evaluating the limits at the very end. The next example illustrates this idea.

EXAMPLE 7.10

Evaluate

$$\int_{\pi/3}^{\pi/2} \frac{\cos x}{\sin x}\ dx.$$

SOLUTION

Since $\cos x$ is the derivative of $\sin x$, it is natural to attempt the substitution $u = \sin x$. Then $du = \cos x\ dx$. [Explain why it would be a bad idea to let $u = \cos x$.] We first treat the improper integral. We find that

$$\int \frac{\cos x}{\sin x}\ dx = \int \frac{du}{u} = \log |u| + C.$$

Now we resubstitute the x-expressions to obtain

$$\int \frac{\cos x}{\sin x}\, dx = \log|\sin x| + C.$$

Finally we can evaluate the original definite integral:

$$\int_{\pi/3}^{\pi/2} \frac{\cos x}{\sin x}\, dx = \log|\sin x|\Big|_{\pi/3}^{\pi/2}$$

$$= \log|\sin \pi/2| - \log|\sin \pi/3| = \log 1 - \log \frac{\sqrt{3}}{2}$$

$$= -\frac{1}{2}\log 3 + \log 2.$$

You Try It: Calculate the integral

$$\int_{-2}^{3} \frac{t\, dt}{(t^2+1)\log(t^2+1)}.$$

7.4 Integrals of Trigonometric Expressions

Trigonometric expressions arise frequently in our work, especially as a result of substitutions. In this section we develop a few examples of trigonometric integrals. The following trigonometric identities will be particularly useful for us.

I We have

$$\sin^2 x = \frac{1 - \cos 2x}{2}.$$

The reason is that

$$\cos 2x = \cos^2 x - \sin^2 x = \left[1 - \sin^2 x\right] - \sin^2 x = 1 - 2\sin^2 x.$$

II We have

$$\cos^2 x = \frac{1 + \cos 2x}{2}.$$

The reason is that

$$\cos 2x = \cos^2 x - \sin^2 x = \cos^2 x - \left[1 - \cos^2 x\right] = 2\cos^2 x - 1.$$

Now we can turn to some examples.

EXAMPLE 7.11

Calculate the integral

$$\int \cos^2 x \, dx.$$

SOLUTION

Of course we will use formula **II**. We write

$$\int \cos^2 x \, dx = \int \frac{1 + \cos 2x}{2} \, dx$$

$$= \int \frac{1}{2} \, dx + \int \frac{1}{2} \cos 2x \, dx$$

$$= \frac{x}{2} + \frac{1}{4} \sin 2x + C.$$

EXAMPLE 7.12

Calculate the integral

$$\int \sin^3 x \cos^2 x \, dx.$$

SOLUTION

When sines and cosines occur together, we always focus on the odd power (when one occurs). We write

$$\sin^3 x \cos^2 x = \sin x \sin^2 x \cos^2 x = \sin x (1 - \cos^2 x) \cos^2 x$$

$$= \left[\cos^2 x - \cos^4 x \right] \sin x.$$

Then

$$\int \sin^3 x \cos^2 \, dx = \int \left[\cos^2 x - \cos^4 x \right] \sin x \, dx.$$

A u-substitution is suggested: We let $u = \cos x$, $du = -\sin x \, dx$. Then the integral becomes

$$-\int \left[u^2 - u^4 \right] du = -\frac{u^3}{3} + \frac{u^5}{5} + C.$$

Resubstituting for the u variable, we obtain the final solution of

$$\int \sin^3 x \cos^2 \, dx = -\frac{\cos^3 x}{3} + \frac{\cos^5 x}{5} + C.$$

You Try It: Calculate the integral

$$\int \sin^2 3x \cos^5 3x \, dx.$$

EXAMPLE 7.13

Calculate

$$\int_0^{\pi/2} \sin^4 x \cos^4 x \, dx.$$

SOLUTION

Substituting

$$\sin^2 x = \frac{1 - \cos 2x}{2} \quad \text{and} \quad \cos^2 x = \frac{1 + \cos 2x}{2}$$

into the integrand yields

$$\int_0^{\pi/2} \left(\frac{1 - \cos 2x}{2}\right)^2 \cdot \left(\frac{1 + \cos 2x}{2}\right)^2 dx$$

$$= \frac{1}{16} \int_0^{\pi/2} 1 - 2\cos^2 2x + \cos^4 2x \, dx.$$

Again using formula **II**, we find that our integral becomes

$$\frac{1}{16} \int_0^{\pi/2} 1 - [1 + \cos 4x] + \left[\frac{1 + \cos 4x}{2}\right]^2 dx$$

$$= \frac{1}{16} \int_0^{\pi/2} 1 - [1 + \cos 4x] + \frac{1}{4}[1 + 2\cos 4x + \cos^2 4x] \, dx.$$

Applying formula **II** one last time yields

$$\frac{1}{16} \int_0^{\pi/2} 1 - [1 + \cos 4x] + \frac{1}{4}\left[1 + 2\cos 4x + \frac{1 + \cos 8x}{2}\right] dx$$

$$= \frac{1}{16}\left[-\frac{1}{4}\sin 4x + \frac{1}{4}\left(x + \frac{1}{2}\sin 4x + \frac{x}{2} + \frac{\sin 8x}{16}\right)\right]_0^{\pi/2}$$

$$= \frac{1}{16}\left(\left[-0 + \frac{1}{4}\left(\frac{\pi}{2} + 0 + \frac{\pi}{4} + 0\right)\right] - \left[-0 + \frac{1}{4}(0 + 0 + 0 + 0)\right]\right)$$

$$= \frac{3\pi}{256}.$$

You Try It: Calculate the integral

$$\int_{\pi/4}^{\pi/3} \sin^3 s \cos^3 s \, ds.$$

You Try It: Calculate the integral

$$\int_{\pi/4}^{\pi/3} \sin^2 s \cos^4 s \, ds.$$

Integrals involving the other trigonometric functions can also be handled with suitable trigonometric identities. We illustrate the idea with some examples that are handled with the identity

$$\tan^2 x + 1 = \frac{\sin^2 x}{\cos^2 x} + 1 = \frac{\sin^2 x + \cos^2 x}{\cos^2 x} = \frac{1}{\cos^2 x} = \sec^2 x.$$

EXAMPLE 7.14

Calculate

$$\int \tan^3 x \sec^3 x \, dx.$$

SOLUTION

Using the same philosophy about odd exponents as we did with sines and cosines, we substitute $\sec^2 x - 1$ for $\tan^2 x$. The result is

$$\int \tan x \left(\sec^2 x - 1\right) \sec^3 x \, dx.$$

We may regroup the terms in the integrand to obtain

$$\int \left[\sec^4 x - \sec^2 x\right] \sec x \tan x \, dx.$$

A u-substitution suggests itself: We let $u = \sec x$ and therefore $du = \sec x \tan x \, dx$. Thus our integral becomes

$$\int u^4 - u^2 \, du = \frac{u^5}{5} - \frac{u^3}{3} + C.$$

Resubstituting the value of u gives

$$\int \tan^3 x \sec^3 x \, dx = \frac{\sec^5 x}{5} - \frac{\sec^3 x}{3} + C.$$

EXAMPLE 7.15

Calculate

$$\int_0^{\pi/4} \sec^4 x \, dx.$$

SOLUTION

We write

$$\int_0^{\pi/4} \sec^4 x \, dx = \int_0^{\pi/4} \sec^2 x \cdot \sec^2 x \, dx$$
$$= \int_0^{\pi/4} (\tan^2 x + 1) \sec^2 x \, dx.$$

Letting $u = \tan x$ and $du = \sec^2 x \, dx$ then gives the integral

$$\int_0^1 u^2 + 1 \, du = \frac{u^3}{3} + u \Big|_0^1$$
$$= \frac{4}{3}.$$

You Try It: Calculate the integral

$$\int_\pi^{2\pi} \sin^6 x \cos^4 x \, dx.$$

Further techniques in the evaluation of trigonometric integrals will be explored in the exercises.

Exercises

1. Use integration by parts to evaluate each of the following indefinite integrals.

(a) $\displaystyle\int \log^2 x \, dx$

(b) $\displaystyle\int x \cdot e^{3x} \, dx$

(c) $\displaystyle\int x^2 \cos x \, dx$

(d) $\displaystyle\int t \sin 3t \cos 3t \, dt$

(e) $\displaystyle\int \cos y \ln(\sin y) \, dy$

(f) $\displaystyle\int x^2 e^{4x} \, dx$

2. Use partial fractions to evaluate each of the following indefinite integrals.

(a) $\displaystyle\int \frac{dx}{(x+2)(x-5)}$

(b) $\displaystyle\int \frac{dx}{(x+1)(x^2+1)}$

(c) $\displaystyle\int \frac{dx}{x^3-2x^2-5x+6}$

(d) $\displaystyle\int \frac{x\,dx}{x^4-1}$

(e) $\displaystyle\int \frac{dx}{x^3-x^2-8x+12}$

(f) $\displaystyle\int \frac{x+1}{x^3-x^2+x-1}$

3. Use the method of u-substitution to evaluate each of the following indefinite integrals.

(a) $\displaystyle\int (1+\sin^2 x)^2 2\sin x \cos x \, dx$

(b) $\displaystyle\int \frac{\sin\sqrt{x}}{\sqrt{x}}\,dx$

(c) $\displaystyle\int \frac{\cos(\ln x)\sin(\ln x)}{x}\,dx$

(d) $\displaystyle\int e^{\tan x}\sec^2 x\,dx$

(e) $\displaystyle\int \frac{\sin x}{1+\cos^2 x}\,dx$

(f) $\displaystyle\int \frac{\sec^2 x}{1-\tan^2 x}\,dx$

4. Evaluate each of the following indefinite trigonometric integrals.

(a) $\displaystyle\int \sin x \cos^2 x \, dx$

(b) $\displaystyle\int \sin^3 x \cos^2 x \, dx$

(c) $\displaystyle\int \tan^3 x \sec^2 x \, dx$

(d) $\displaystyle\int \tan x \sec^3 x \, dx$

(e) $\displaystyle\int \sin^2 x \cos^2 x \, dx$

(f) $\displaystyle\int \sin x \cos^4 x \, dx$

5. Calculate each of the following definite integrals.

(a) $\displaystyle\int_0^1 e^x \sin x \, dx$

(b) $\displaystyle\int_1^e x^2 \ln x \, dx$

(c) $\displaystyle\int_2^4 \frac{(2x+1) \, dx}{x^3 + x^2}$

(d) $\displaystyle\int_0^\pi \sin^2 x \cos^2 x \, dx$

(e) $\displaystyle\int_{\pi/4}^{\pi/3} \tan x \sec x \, dx$

(f) $\displaystyle\int_0^{\pi/4} \frac{\tan x}{\cos x} \, dx$

CHAPTER 8

Applications of the Integral

8.1   Volumes by Slicing

8.1.0   INTRODUCTION

When we learned the theory of the integral, we found that the basic idea was that one can calculate the area of an irregularly shaped region by subdividing the region into "rectangles." We put the word "rectangle" here in quotation marks because the region is not *literally* broken up into rectangles; the union of the rectangles differs from the actual region under consideration by some small errors (see Fig. 8.1). But the contribution made by these errors vanishes as the mesh of the rectangles become finer and finer.

   We will now implement this same philosophy to calculate certain volumes. Some of these will be volumes that you have heard about (e.g., the sphere or cone), but have never known *why* the volume had the value that it had. Others will be entirely new (e.g., the paraboloid of revolution). We will again use the method of slicing.

8.1.1   THE BASIC STRATEGY

Imagine a solid object situated as in Fig. 8.2. Observe the axes in the diagram, and imagine that we slice the figure with slices that are vertical (i.e., that rise out of the x-y plane) and that are perpendicular to the x-axis (and parallel to the y-axis). Look at Fig. 8.3. Notice, in the figure, that the figure extends from $x = a$ to $x = b$.

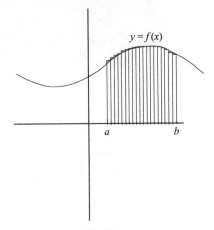

Fig. 8.1

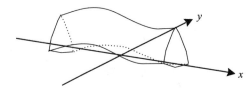

Fig. 8.2

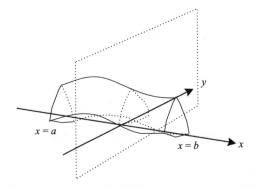

Fig. 8.3

If we can express the *area* of the slice at position x as a function $A(x)$ of x, then (see Fig. 8.4) the *volume* of a slice of thickness $\triangle x$ at position x will be about $A(x) \cdot \triangle x$. If $\mathcal{P} = \{x_0, x_1, \ldots, x_k\}$ is a partition of the interval $[a, b]$ then the volume of the original solid object will be about

$$\widetilde{V} = \sum_j A(x_j) \cdot \triangle x.$$

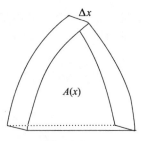

Fig. 8.4

As the mesh of the partition becomes finer and finer, this (Riemann) sum will tend to the integral

$$\int_a^b A(x)\,dx.$$

We declare the value of this integral to be the *volume V of the solid object.*

8.1.2 EXAMPLES

EXAMPLE 8.1

Calculate the volume of the right circular cone with base a disc of radius 3 and height 6.

SOLUTION

Examine Fig. 8.5. We have laid the cone on its side, so that it extends from $x = 0$ to $x = 6$. The upper edge of the figure is the line $y = 3 - x/2$. At position x,

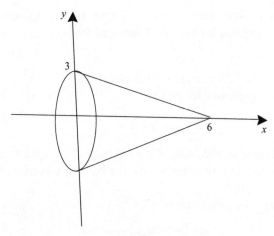

Fig. 8.5

the height of the upper edge is $3 - x/2$, and that number is also the radius of the circular slice at position x (Fig. 8.6). Thus the area of that slice is

$$A(x) = \pi \left(3 - \frac{x}{2}\right)^2.$$

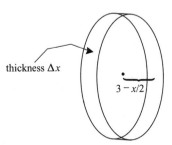

thickness Δx

$3 - x/2$

Fig. 8.6

We find then that the volume we seek is

$$V = \int_0^6 A(x)\,dx = \int_0^6 \pi \left(3 - \frac{x}{2}\right)^2 dx = -\pi \frac{2(3 - x/2)^3}{3}\bigg|_0^6 = 18\pi.$$

You Try It: Any book of tables (see [CRC]) will tell you that the volume of a right circular cone of base radius r and height h is $\frac{1}{3}\pi r^2 h$. This formula is consistent with the result that we obtained in the last example for $r = 3$ and $h = 6$. Use the technique of Example 8.1 to verify this more general formula.

EXAMPLE 8.2

A solid has base the unit disk in the x-y plane. The vertical cross section at position x is an equilateral triangle. Calculate the volume.

SOLUTION

Examine Fig. 8.7. The unit circle has equation $x^2 + y^2 = 1$. For our purposes, this is more conveniently written as

$$y = \pm\sqrt{1 - x^2}. \tag{$\star$}$$

Thus the endpoints of the base of the equilateral triangle at position x are the points $(x, \pm\sqrt{1 - x^2})$. In other words, the base of this triangle is

$$b = 2\sqrt{1 - x^2}.$$

Examine Fig. 8.8. We see that an equilateral triangle of side b has height $\sqrt{3}b/2$. Thus the area of the triangle is $\sqrt{3}b^2/4$. In our case then, the equilateral

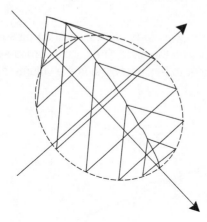

Fig. 8.7

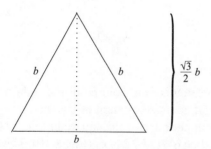

Fig. 8.8

triangular slice at position x has area

$$A(x) = \frac{\sqrt{3}}{4} \cdot \left[2\sqrt{1 - x^2}\right]^2 = \sqrt{3}(1 - x^2).$$

Finally, we may conclude that the volume we seek is

$$V = \int_{-1}^{1} A(x) \, dx$$

$$= \int_{-1}^{1} \sqrt{3}(1 - x^2) \, dx$$

$$= \sqrt{3}\left[x - \frac{x^3}{3}\right]_{-1}^{1}$$

$$= \sqrt{3}\left[\left(1 - \frac{1}{3}\right) - \left((-1) - \frac{-1}{3}\right)\right]$$

$$= \frac{4\sqrt{3}}{3}.$$

EXAMPLE 8.3

A solid has base in the *x-y* plane consisting of a unit square with center at the origin and vertices on the axes. The vertical cross-section at position *x* is itself a square. Refer to Fig. 8.9. What is the volume of this solid?

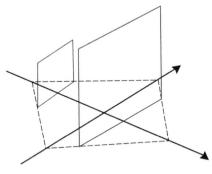

Fig. 8.9

SOLUTION

It is sufficient to calculate the volume of the right half of this solid, and to double the answer. Of course the extent of x is then $0 \leq x \leq 1/\sqrt{2}$. At position x, the height of the upper edge of the square base is $1/\sqrt{2} - x$. So the base of the vertical square slice is $2(1/\sqrt{2} - x)$ (Fig. 8.10). The area of the slice is then

$$A(x) = \left[2(1/\sqrt{2} - x)\right]^2 = \left(\sqrt{2} - 2x\right)^2.$$

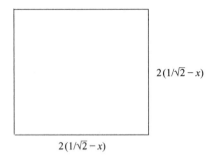

Fig. 8.10

It follows that

$$V = 2 \cdot \int_0^{1/\sqrt{2}} A(x)\,dx$$

$$= 2 \int_0^{1/\sqrt{2}} \left(\sqrt{2} - 2x\right)^2 dx$$

$$= 2 \left[-\frac{\left(\sqrt{2} - 2x\right)^3}{6} \right]_0^{1/\sqrt{2}}$$

$$= 2 \left[-\frac{0^3}{6} - \left(-\frac{2\sqrt{2}}{6} \right) \right]$$

$$= \frac{2\sqrt{2}}{3}.$$

You Try It: Calculate the volume of the solid with base in the plane an equilateral triangle of side 1, with base on the x-axis, and with vertical cross-section parallel to the y-axis consisting of an equilateral triangle.

EXAMPLE 8.4

Calculate the volume inside a sphere of radius 1.

SOLUTION

It is convenient for us to think of the sphere as centered at the origin in the x-y plane. Thus (Fig. 8.11) the slice at position x, $-1 \le x \le 1$, is a disk. Since we are working with base the unit circle, we may calculate (just as in Example 8.2) that the diameter of this disk is $2\sqrt{1 - x^2}$. Thus the radius is $\sqrt{1 - x^2}$ and the area is

$$A(x) = \pi \cdot \left(\sqrt{1 - x^2}\right)^2 = \pi \cdot (1 - x^2).$$

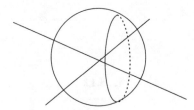

Fig. 8.11

In conclusion, the volume we seek is

$$V = \int_{-1}^{1} \pi(1 - x^2) \, dx.$$

We easily evaluate this integral as follows:

$$V = \pi \cdot \left[x - \frac{x^3}{3} \right]_{-1}^{1}$$

$$= \pi \cdot \left[\left(1 - \frac{1}{3} \right) - \left(-1 - \frac{-1}{3} \right) \right]$$

$$= \frac{4}{3}\pi.$$

You Try It: Any book of tables (see [CRC]) will tell you that the volume inside a sphere of radius r is $4\pi r^3/3$. This formula is consistent with the answer we obtained in the last example for $r = 1$. Use the method of this section to derive this more general formula for arbitrary r.

8.2   Volumes of Solids of Revolution

8.2.0   INTRODUCTION

A useful way—and one that we encounter frequently in everyday life—for generating solids is by revolving a planar region about an axis. For example, we can think of a ball (the interior of a sphere) as the solid obtained by rotating a disk about an axis (Fig. 8.12). We can think of a cylinder as the solid obtained by rotating a rectangle about an adjacent axis (Fig. 8.13). We can think of a tubular solid as obtained by rotating a rectangle around a non-adjacent axis (Fig. 8.14).

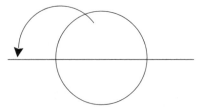

Fig. 8.12

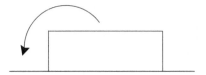

Fig. 8.13

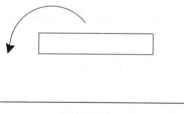

Fig. 8.14

There are two main methods for calculating volumes of solids of revolution: the method of washers and the method of cylinders. The first of these is really an instance of volume by slicing, just as we saw in the last section. The second uses a different geometry; instead of slices one uses cylindrical shells. We shall develop both technologies by way of some examples.

8.2.1   THE METHOD OF WASHERS

EXAMPLE 8.5

A solid is formed by rotating the triangle with vertices (0, 0), (2, 0), and (1, 1) about the *x*-axis. See Fig. 8.15. What is the resulting volume?

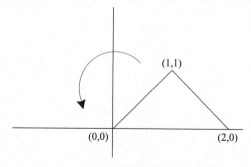

Fig. 8.15

SOLUTION

For $0 \leq x \leq 1$, the upper edge of the triangle has equation $y = x$. Thus the segment being rotated extends from $(x, 0)$ to (x, x). Under rotation, it will generate a disk of radius x, and hence area $A(x) = \pi x^2$. Thus the volume generated over the segment $0 \leq x \leq 1$ is

$$V_1 = \int_0^1 \pi x^2 \, dx.$$

Similarly, for $1 \leq x \leq 2$, the upper edge of the triangle has equation $y = 2 - x$. Thus the segment being rotated extends from $(x, 0)$ to $(x, 2 - x)$. Under rotation, it will generate a disk of radius $2-x$, and hence area $A(x) = \pi(2-x)^2$. Thus the volume generated over the segment $1 \le x \le 2$ is

$$V_2 = \int_1^2 \pi(2-x)^2 \, dx.$$

In summary, the total volume of our solid of revolution is

$$V = V_1 + V_2$$

$$= \pi \left[\frac{x^3}{3} \Big|_0^1 + \frac{-(2-x)^3}{3} \Big|_1^2 \right]$$

$$= \pi \left(\frac{1}{3} - 0 \right) + \left(-0 - \left[-\frac{1}{3} \right] \right)$$

$$= \frac{2\pi}{3}.$$

EXAMPLE 8.6

The portion of the curve $y = x^2$ between $x = 1$ and $x = 4$ is rotated about the x-axis (Fig. 8.16). What volume does the resulting surface enclose?

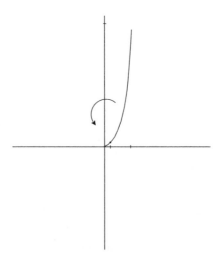

Fig. 8.16

SOLUTION

At position x, the curve is x^2 units above the x-axis. The point (x, x^2), under rotation, therefore generates a circle of radius x^2. The disk that the circle bounds has area $A(x) = \pi \cdot (x^2)^2$. Thus the described volume is

$$V = \int_1^4 \pi \cdot x^4 \, dx = \pi \cdot \left. \frac{x^5}{5} \right|_1^4 = \frac{1023\pi}{5}.$$

Math Note: The reasoning we have used in the last two examples shows this: If the curve $y = f(x)$, $a \leq x \leq b$, is rotated about the x-axis then the volume enclosed by the resulting surface is

$$V = \int_a^b \pi \cdot [f(x)]^2 \, dx.$$

You Try It: Calculate the volume enclosed by the surface obtained by rotating the curve $y = \sqrt{x} + 1$, $4 \leq x \leq 9$, about the x-axis.

EXAMPLE 8.7

The curve $y = x^3$, $0 \leq x \leq 3$, is rotated about the y-axis. What volume does the resulting surface enclose?

SOLUTION

It is convenient in this problem to treat y as the independent variable and x as the dependent variable. So we write the curve as $x = y^{1/3}$. Then, at position y, the curve is distance $y^{1/3}$ from the axis so the disk generated under rotation will have radius $y^{1/3}$ (Fig. 8.17). Thus the disk will have area $A(y) = \pi \cdot [y^{1/3}]^2$. Also, since x ranges from 0 to 3 we see that y ranges from 0 to 27. As a result,

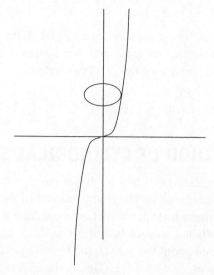

Fig. 8.17

the volume enclosed is

$$V = \int_0^{27} \pi \cdot y^{2/3} \, dy = \pi \cdot \frac{y^{5/3}}{5/3} \Big|_0^{27} = \frac{729\pi}{5}.$$

Math Note: The reasoning we have used in the last example shows this: If the curve $x = g(y)$, $c \leq y \leq d$, is rotated about the y-axis then the volume enclosed by the resulting surface is

$$V = \int_c^d \pi \cdot [g(y)]^2 \, dy.$$

You Try It: Calculate the volume enclosed when the curve $y = x^{1/3}$, $32 \leq x \leq 243$, is rotated about the y-axis.

EXAMPLE 8.8

Set up, but do not evaluate, the integral that represents the volume generated when the planar region between $y = x^2 + 1$ and $y = 2x + 4$ is rotated about the x-axis.

SOLUTION

When the planar region is rotated about the x-axis, it will generate a donut-shaped solid. Notice that the curves intersect at $x = -1$ and $x = 3$; hence the intersection lies over the interval $[-1, 3]$. For each x in that interval, the segment connecting $(x, x^2 + 1)$ to $(x, 2x + 4)$ will be rotated about the x-axis. *It will generate a washer.* See Fig. 8.18. The area of that washer is

$$A(x) = \pi \cdot [2x + 4]^2 - \pi \cdot [x^2 + 1].$$

[*Notice that we calculate the area of a washer by subtracting the areas of two circles—not by subtracting the radii and then squaring.*]

It follows that the volume of the solid generated is

$$V = \int_{-1}^3 \pi \cdot [2x + 4]^2 - \pi \cdot [x^2 + 1] \, dx.$$

8.2.2 THE METHOD OF CYLINDRICAL SHELLS

Our philosophy will now change. When we divide our region up into vertical strips, we will now rotate each strip about the y-axis instead of the x-axis. Thus, instead of generating a disk with each strip, we will now generate a cylinder.

Look at Fig. 8.19. When a strip of height h and thickness $\triangle x$, with distance r from the y-axis, is rotated about the y-axis, the resulting cylinder has surface area $2\pi r \cdot h$ and *volume* about $2\pi r \cdot h \cdot \triangle x$. This is the expression that we will treat in order to sum up the volumes of the cylinders.

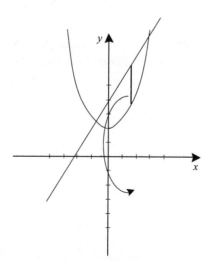

Fig. 8.18

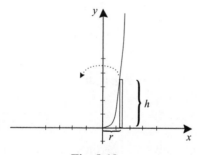

Fig. 8.19

EXAMPLE 8.9

Use the method of cylindrical shells to calculate the volume of the solid enclosed when the curve $y = x^2$, $1 \le x \le 3$, is rotated about the y-axis.

SOLUTION

As usual, we think of the region under $y = x^2$ and above the x-axis as composed of vertical segments or strips. The segment at position x has height x^2. Thus, in this instance, $h = x^2$, $r = x$, and the volume of the cylinder is $2\pi x \cdot x^2 \cdot \Delta x$. As a result, the requested volume is

$$V = \int_1^3 2\pi x \cdot x^2 \, dx.$$

We easily calculate this to equal

$$V = 2\pi \cdot \int_1^3 x^3 \, dx = 2\pi \frac{x^4}{4} \Big|_1^3 = 2\pi \left[\frac{3^4}{4} - \frac{1^4}{4} \right] = 40\pi.$$

EXAMPLE 8.10

Use the method of cylindrical shells to calculate the volume enclosed when the curve $y = x^2$, $0 \leq x \leq 3$, is rotated about the x-axis (Fig. 8.20).

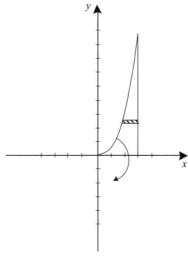

Fig. 8.20

SOLUTION

We reverse, in our analysis, the roles of the x- and y-axes. Of course y ranges from 0 to 9. For each position y in that range, there is a segment stretching from $x = \sqrt{y}$ to $x = 3$. Thus it has length $3 - \sqrt{y}$. Then the cylinder generated when this segment (thickened to a strip of width Δy) is rotated about the x-axis has volume

$$V(y) = 2\pi y \cdot \left[3 - \sqrt{y}\right] \Delta y.$$

The aggregate volume is then

$$V = \int_0^9 2\pi y \cdot \left[3 - \sqrt{y}\right] dy$$

$$= 2\pi \cdot \int_0^9 3y - y^{3/2} \, dy$$

$$= 2\pi \cdot \left[\frac{3y^2}{2} - \frac{y^{5/2}}{5/2}\right]_0^9 dy$$

$$= 2\pi \cdot \left[\left(\frac{243}{2} - \frac{2 \cdot 243}{5}\right) - \left(\frac{0}{2} - \frac{0}{5}\right)\right]$$

$$= 2\pi \cdot \frac{243}{10}$$

$$= \frac{243\pi}{5}.$$

You Try It: Use the method of cylindrical shells to calculate the volume enclosed when the region $0 \le y \le \sin x$, $0 \le x \le \pi/2$, is rotated about the y-axis.

8.2.3   DIFFERENT AXES

Sometimes it is convenient to rotate a curve about some line other than the coordinate axes. We now provide a couple of examples of that type of problem.

EXAMPLE 8.11

Use the method of washers to calculate the volume of the solid enclosed when the curve $y = \sqrt{x}$, $1 \le x \le 4$, is rotated about the line $y = -1$. See Fig. 8.21.

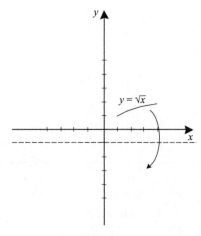

Fig. 8.21

SOLUTION

The key is to notice that, at position x, the segment to be rotated has height $\sqrt{x} - (-1)$—the distance from the point $(x, \sqrt{x})$ on the curve to the line $y = -1$. Thus the disk generated has area $A(x) = \pi \cdot (\sqrt{x} + 1)^2$. The resulting aggregate volume is

$$V = \int_1^4 \pi \cdot \left(\sqrt{x} + 1\right)^2 dx$$

$$= \pi \int_1^4 x + 2\sqrt{x} + 1 \, dx$$

$$= \pi \left[\frac{x^2}{2} + \frac{2x^{3/2}}{3/2} + x \right]_1^4$$

$$= \pi \cdot \left[\frac{4^2}{2} + \frac{2 \cdot 8}{3/2} + 4 \right] - \pi \cdot \left[\frac{1^2}{2} + \frac{2 \cdot 1}{3/2} + 1 \right]$$

$$= \frac{119}{6} \pi.$$

You Try It: Calculate the volume inside the surface generated when $y = \sqrt[3]{x} + x$ is rotated about the line $y = -3$, $1 \le x \le 4$.

EXAMPLE 8.12

Calculate the volume of the solid enclosed when the area between the curves $x = (y - 2)^2 + 1$ and $x = -(y - 2)^2 + 9$ is rotated about the line $y = -2$.

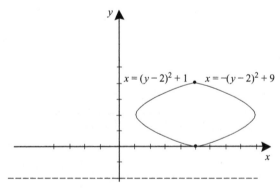

Fig. 8.22

SOLUTION

Solving the equations simultaneously, we find that the points of intersection are $(5, 0)$ and $(5, 4)$. The region between the two curves is illustrated in Fig. 8.22.

At height y, the horizontal segment that is to be rotated stretches from $((y - 2)^2 + 1, y)$ to $(-(y - 2)^2 + 9, y)$. Thus the cylindrical shell that is generated has radius $y - 2$, height $8 - 2(y - 2)^2$, and thickness $\triangle y$. It therefore generates the element of volume given by

$$2\pi \cdot (y - 2) \cdot [8 - 2(y - 2)^2] \cdot \triangle y.$$

The aggregate volume that we seek is therefore

$$V = \int_0^4 2\pi \cdot (y - 2) \cdot [8 - 2(y - 2)^2] \, dy$$

$$= \int_0^4 16\pi (y - 2) - 4\pi (y - 2)^3 \, dy$$

$$= 8\pi (y - 2)^2 - \pi (y - 4)^4 \big|_0^4$$

$$= 256\pi.$$

You Try It: Calculate the volume enclosed when the curve $y = \cos x$ is rotated about the line $y = 4$, $\pi \le x \le 3\pi$.

8.3   Work

One of the basic principles of physics is that work performed is force times distance: If you apply force F pounds in moving an object d feet, then the work is

$$W = F \cdot d \quad \text{foot-pounds.}$$

The problem becomes more interesting (than simple arithmetic) if the force is varying from point to point. We now consider some problems of that type.

EXAMPLE 8.13

A weight is pushed in the plane from $x = 0$ to $x = 10$. Because of a prevailing wind, the force that must be applied at point x is $F(x) = 3x^2 - x + 10$ foot-pounds. What is the total work performed?

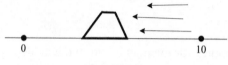

Fig. 8.23

SOLUTION
   Following the way that we usually do things in calculus, we break the problem up into pieces. In moving the object from position x to position $x + \triangle x$, the

distance moved is Δx feet and the force applied is about $F(x) = 3x^2 - x + 10$. See Fig. 8.23. Thus work performed in that little bit of the move is $w(x) = (3x^2 - x + 10) \cdot \Delta x$. The aggregate of the work is obtained by summation. In this instance, the integral is the appropriate device:

$$W = \int_0^{10} 3x^2 - x + 10 \, dx = x^3 - \frac{x^2}{2} + 10x \Big|_0^{10} = 1050 \text{ foot-pounds}.$$

EXAMPLE 8.14

A man is carrying a 100 lb sack of sand up a 20-foot ladder at the rate of 5 feet per minute. The sack has a hole in it and sand leaks out continuously at a rate of 4 lb per minute. How much work does the man do in carrying the sack?

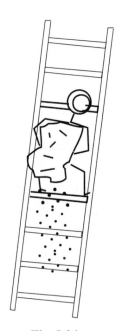

Fig. 8.24

SOLUTION

It takes four minutes for the man to climb the ladder. At time t, the sack has $100 - 4t$ pounds of sand in it. From time t to time $t + \Delta t$, the man moves $5 \cdot \Delta t$ feet up the ladder. He therefore performs about $w(t) = (100 - 4t) \cdot 5 \Delta t$ foot-pounds of work. See Fig. 8.24. The total work is then the integral

$$W = \int_0^4 (100 - 4t) \, 5 \, dt = 500t - 10t^2 \Big|_0^4 = 1840 \text{ foot-pounds}.$$

You Try It: A man drags a 100 pound weight from $x = 0$ to $x = 300$. He resists a wind which at position x applies a force of magnitude $F(x) = x^3 + x + 40$. How much work does he perform?

EXAMPLE 8.15

According to Hooke's Law, the amount of force exerted by a spring is proportional to the distance of its displacement from the rest position. The constant of proportionality is called the *Hooke's constant*. A certain spring exerts a force of 10 pounds when stretched 1/2 foot beyond its rest state. What is the work performed in stretching the spring from rest to 1/3 foot beyond its rest length?

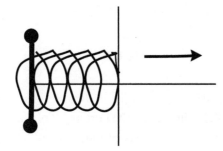

Fig. 8.25

SOLUTION

Let the x-variable denote the position of the right end of the spring (Fig. 8.25), with $x = 0$ the rest position. The left end of the spring is pinned down. Imagine that the spring is being stretched to the right. We know that the force exerted by the spring has the form

$$F(x) = kx,$$

with k a negative constant (since the spring will pull to the left). Also $F(0.5) = -10$. It follows that $k = -20$, so that

$$F(x) = -20x.$$

Now the work done in moving the spring from position x to position $x + \Delta x$ will be about $(20x) \cdot \Delta x$ (the sign is $+$ since we will pull the spring to the right—*against* the pull of the spring). Thus the total work done in stretching the right end of the spring from $x = 0$ to $x = 1/3$ is

$$W = \int_0^{1/3} (20x)\, dx = 10x^2 \Big|_0^{1/3} = \frac{10}{9} \text{ foot-pounds.}$$

EXAMPLE 8.16

Imagine that a water tank in the shape of a hemisphere of radius 10 feet is being pumped out (Fig. 8.26). Find the work done in lowering the water level from 1 foot from the top of the tank to 3 feet from the top of the tank.

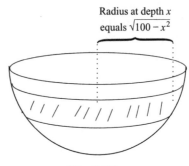

Radius at depth x equals $\sqrt{100 - x^2}$

Fig. 8.26

SOLUTION

A glance at Fig. 8.27 shows that the horizontal planar slice of the tank, at the level x feet below the top, is a disk of radius $\sqrt{100 - x^2}$. This disk therefore has area $A(x) = \pi \cdot (100 - x^2)$. Thus a slice at that level of thickness $\triangle x$ will have volume

$$V(x) = \pi \cdot (100 - x^2) \cdot \triangle x$$

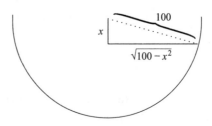

x 100 $\sqrt{100 - x^2}$

Fig. 8.27

and (assuming that water weights 62.4 pounds per cubic foot) weight equal to

$$w(x) = 62.4\pi \cdot (100 - x^2) \cdot \triangle x.$$

Thus the work in raising this slice to the top of the tank (where it can then be dumped) is

$$W(x) = \left[62.4\pi \cdot (100 - x^2) \cdot \triangle x \right] \cdot x \text{ foot-pounds.}$$

We calculate the total work by adding all these elements together using an integral. The result is

$$W = \int_1^3 \left[62.4\pi \cdot (100 - x^2) \cdot x \right] dx$$

$$= 62.4\pi \cdot \int_1^3 100x - x^3 \, dx$$

$$= 62.4\pi \left[50x^2 - \frac{x^4}{4} \right]_1^3$$

$$= 62.4\pi \left[\left(450 - \frac{81}{4} \right) - \left(50 - \frac{1}{4} \right) \right]$$

$$= 23{,}712\pi \text{ foot-pounds.}$$

You Try It: A spring has Hooke's constant 5. How much work is performed in stretching the spring half a foot from its rest position?

8.4   Averages

In ordinary conversation, when we average a collection $p_1, \ldots, p_k$ of k numbers, we add them together and divide by the number of items:

$$\sigma = \text{Average} = \frac{p_1 + \cdots + p_k}{k}.$$

The significance of the number σ is that if we wanted all the k numbers to be equal, but for the total to be the same, then that common value would have to be σ.

Now suppose that we want to average a continuous function f over an interval $[a, b]$ of its domain. We can partition the interval,

$$\mathcal{P} = \{x_0, x_1, \ldots, x_k\},$$

with $x_0 = a$ and $x_k = b$ as usual. We assume that this is a *uniform partition*, with $x_j - x_{j-1} = \Delta x = (b-a)/k$ for all j. Then an "approximate average" of f would be given by

$$\sigma_{\text{app}} = \frac{f(x_1) + f(x_2) + \cdots + f(x_k)}{k}.$$

It is convenient to write this expression as

$$\sigma_{\text{app}} = \frac{1}{b-a} \sum_{j=1}^k f(x_j) \cdot \frac{b-a}{k} = \frac{1}{b-a} \sum_{j=1}^k f(x_j) \cdot \Delta x.$$

This last is a Riemann sum for the integral $(1/[b-a]) \cdot \int_a^b f(x)\,dx$. Thus, letting the mesh of the partition go to zero, we declare

$$\text{average of } f = \sigma = \frac{1}{b-a} \int_a^b f(x)\,dx.$$

EXAMPLE 8.17

In a tropical rain forest, the rainfall at time t is given by $\varphi(t) = 0.1 - 0.1t + 0.05t^2$ inches per hour, $0 \le t \le 10$. What is the average rainfall for times $0 \le t \le 6$?

SOLUTION

We need only average the function φ:

$$\text{average rainfall} = \sigma = \frac{1}{6-0} \int_0^6 \varphi(t)\,dt$$

$$= \frac{1}{6} \int_0^6 0.1 - 0.1t + 0.05t^2\,dt$$

$$= \frac{1}{6}\left[0.1t - 0.05t^2 + \frac{0.05}{3}t^3 \right]_0^6$$

$$= 0.1 - 0.3 + 0.6$$

$$= 0.4 \text{ inches per hour.}$$

EXAMPLE 8.18

Let $f(x) = x/2 - \sin x$ on the interval $[-2, 5]$. Compare the average value of this function on the interval with its minimum and maximum.

SOLUTION

Observe that

$$f'(x) = \frac{1}{2} - \cos x.$$

Thus the critical points occur when $\cos x = 1/2$, or at $-\pi/3, \pi/3$. We also must consider the endpoints $-2, 5$. The values at these points are

$$f(-2) = -1 + \sin 2 \approx -0.0907026$$

$$f(-\pi/3) = -\frac{\pi}{6} + \frac{\sqrt{3}}{2} \approx 0.3424266$$

$$f(\pi/3) = \frac{\pi}{6} - \frac{\sqrt{3}}{2} \approx -0.3424266$$

$$f(5) = \frac{5}{2} - \sin 5 \approx 3.458924.$$

Plainly, the maximum value is $f(5) = 5/2 - \sin 5 \approx 3.458924$. The minimum value is $f(\pi/3) \approx -0.3424266$.

The average value of our function is

$$
\begin{aligned}
\sigma &= \frac{1}{5 - (-2)} \int_{-2}^{5} \frac{x}{2} - \sin x \, dx \\
&= \frac{1}{7} \left[\frac{x^2}{4} + \cos x \right]_{-2}^{5} \\
&= \frac{1}{7} \left[\left(\frac{25}{4} + \cos 5 \right) - \left(\frac{4}{4} + \cos 2 \right) \right] \\
&= \frac{1}{7} \left[\frac{21}{4} + \cos 5 - \cos 2 \right] \\
&\approx 0.84997.
\end{aligned}
$$

You can see that the average value lies between the maximum and the minimum, as it should. This is an instance of a general phenomenon.

You Try It: On a certain tree line, the height of trees at position x is about $100 - 3x + \sin 5x$. What is the average height of trees from $x = 2$ to $x = 200$?

EXAMPLE 8.19
What is the average value of the function $g(x) = \sin x$ over the interval $[0, 2\pi]$?

SOLUTION
We calculate that

$$
\sigma = \frac{1}{2\pi - 0} \int_{0}^{2\pi} \sin x \, dx = \frac{1}{2\pi} [-\cos x] \Big|_{0}^{2\pi} = \frac{1}{2\pi} [-1 - (-1)] = 0.
$$

We see that this answer is consistent with our intuition: the function $g(x) = \sin x$ takes positive values and negative values with equal weight over the interval $[0, 2\pi]$. The *average* is intuitively equal to zero. And that is the actual computed value.

You Try It: Give an example of a function on the real line whose average over *every* interval of length 4 is 0.

8.5   Arc Length and Surface Area

Just as the integral may be used to calculate planar area and spatial volume, so this tool may also be used to calculate the arc length of a curve and surface area. The basic idea is to approximate the length of a curve by the length of its piecewise linear approximation. A similar comment applies to the surface area. We begin by describing the basic rubric.

8.5.1   ARC LENGTH

Suppose that $f(x)$ is a function on the interval $[a, b]$. Let us see how to calculate the length of the curve consisting of the graph of f over this interval (Fig. 8.28). We partition the interval:

$$a = x_0 \leq x_1 \leq x_2 \leq \cdots \leq x_{k-1} \leq x_k = b.$$

Look at Fig. 8.29. Corresponding to each pair of points x_{j-1}, x_j in the partition is a segment connecting two points on the curve; the segment has endpoints $(x_{j-1}, f(x_{j-1}))$ and $(x_j, f(x_j))$. The length ℓ_j of this segment is given by the

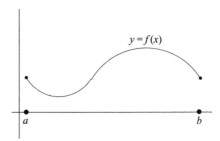

Fig. 8.28

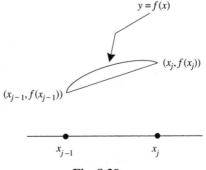

Fig. 8.29

usual planar distance formula:

$$\ell_j = \left([x_j - x_{j-1}]^2 + [f(x_j) - f(x_{j-1})]^2\right)^{1/2}.$$

We denote the quantity $x_j - x_{j-1}$ by Δx and apply the definition of the derivative to obtain

$$\frac{f(x_j) - f(x_{j-1})}{\Delta x} \approx f'(x_j).$$

Now we may rewrite the formula for ℓ_j as

$$\ell_j \approx ([\Delta x]^2 + [f'(x_j)\Delta x]^2)^{1/2}$$
$$= (1 + [f'(x_j)]^2)^{1/2}\Delta x.$$

Summing up the lengths ℓ_j (Fig. 8.30) gives an approximate length for the curve:

$$\text{length of curve} \approx \sum_{j=1}^{k}\ell_j = \sum_{j=1}^{k}(1 + [f'(x_j)]^2)^{1/2}\Delta x.$$

Fig. 8.30

But this last is a Riemann sum for the integral

$$\ell = \int_a^b (1 + [f'(x)]^2)^{1/2}\, dx. \qquad\qquad (\star)$$

As the mesh of the partition becomes finer, the approximating sum is ever closer to what we think of as the length of the curve, and it also converges to this integral. Thus the integral represents the length of the curve.

EXAMPLE 8.20

Let us calculate the arc length of the graph of $f(x) = 4x^{3/2}$ over the interval $[0, 3]$.

SOLUTION

The length is

$$\int_0^3 (1 + [f'(x)]^2)^{1/2}\, dx = \int_0^3 (1 + [6x^{1/2}]^2)^{1/2}\, dx$$

$$= \int_0^3 (1 + 36x)^{1/2}\, dx$$

$$= \frac{1}{54} \cdot (1 + 36x)^{3/2}\Big|_0^3$$

$$= \frac{1}{54}[109^{3/2} - 1^{3/2}]$$

$$= \frac{(109)^{3/2} - 1}{54}.$$

EXAMPLE 8.21

Let us calculate the length of the graph of the function $f(x) = (1/2)\times (e^x + e^{-x})$ over the interval $[1, \ln 8]$.

SOLUTION

We calculate that

$$f'(x) = (1/2)(e^x - e^{-x}).$$

Therefore the length of the curve is

$$\int_1^{\ln 8} \left(1 + [(1/2)(e^x - e^{-x})]^2\right)^{1/2}\, dx$$

$$= \int_1^{\ln 8} \left(\frac{e^{2x}}{4} + \frac{1}{2} + \frac{e^{-2x}}{4}\right)^{1/2}\, dx$$

$$= \frac{1}{2}\int_1^{\ln 8} e^x + e^{-x}\, dx$$

$$= \frac{1}{2}[e^x - e^{-x}]_1^{\ln 8}$$

$$= \frac{63}{16} - \frac{e}{2} + \frac{1}{2e}.$$

You Try It: Set up, but do not evaluate, the integral for the arc length of the graph of $y = \sqrt{\sin x}$ on the interval $\pi/4 \le x \le 3\pi/4$.

Sometimes an arc length problem is more conveniently solved if we think of the curve as being the graph of $x = g(y)$. Here is an example.

EXAMPLE 8.22

Calculate the length of that portion of the graph of the curve $16x^2 = 9y^3$ between the points $(0, 0)$ and $(6, 4)$.

SOLUTION

We express the curve as

$$x = \frac{3}{4}y^{3/2}, \qquad 0 \leq y \leq 4.$$

Then $dx/dy = \frac{9}{8}y^{1/2}$. Now, reversing the roles of x and y in $(\star)$, we find that the requested length is

$$\int_0^4 \sqrt{1 + [(9/8)y^{1/2}]^2} \, dy = \int_0^4 \sqrt{1 + (81/64)y} \, dy.$$

This integral is easily evaluated and we see that it has value $[2 \cdot (97)^{3/2} - 128]/243$.

Notice that the last example would have been considerably more difficult (the integral would have been harder to evaluate) had we expressed the curve in the form $y = f(x)$.

You Try It: Write the integral that represents the length of a semi-circle and evaluate it.

8.5.2   SURFACE AREA

Let $f(x)$ be a non-negative function on the interval $[a, b]$. Imagine rotating the graph of f about the x-axis. This procedure will generate a surface of revolution, as shown in Fig. 8.31. We will develop a procedure for determining the area of such a surface.

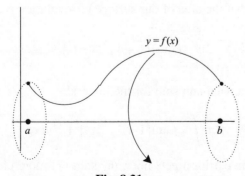

Fig. 8.31

We partition the interval $[a, b]$:

$$a = x_0 \leq x_1 \leq x_2 \leq \cdots \leq x_{k-1} \leq x_k = b.$$

Corresponding to each pair of elements x_{j-1}, x_j in the partition is a portion of curve,

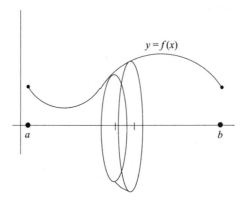

Fig. 8.32

as shown in Fig. 8.32. When that piece of curve is rotated about the x-axis, we obtain a cylindrical surface. Now the area of a true right circular cylinder is $2\pi \cdot r \cdot h$. We do not have a true cylinder, so we proceed as follows. We may approximate the radius by $f(x_j)$. And the height of the cylinder can be approximated by the length of the curve spanning the pair x_{j-1}, x_j. This length was determined above to be about

$$\left(1 + [f'(x_j)]^2\right)^{1/2} \Delta x_j.$$

Thus the area contribution of this cylindrical increment of our surface is about

$$2\pi \cdot f(x_j)\left(1 + [f'(x_j)]^2\right)^{1/2} \Delta x_j.$$

See Fig. 8.33. If we sum up the area contribution from each subinterval of the partition we obtain that the area of our surface of revolution is about

$$\sum_{j=1}^{k} 2\pi \cdot f(x_j)\left(1 + [f'(x_j)]^2\right)^{1/2} \Delta x_j. \qquad (*)$$

But this sum is also a Riemann sum for the integral

$$2\pi \int_a^b f(x)\left(1 + [f'(x)]^2\right)^{1/2} dx.$$

As the mesh of the partition gets finer, the sum $(*)$ more closely approximates what we think of as the area of the surface, but it also converges to the integral.

Δx

Fig. 8.33

We conclude that the integral

$$2\pi \int_a^b f(x)(1 + [f'(x)]^2)^{1/2}\, dx$$

represents the area of the surface of revolution.

EXAMPLE 8.23

Let $f(x) = 2x^3$. For $1 \le x \le 2$ we rotate the graph of f about the x-axis. Calculate the resulting surface area.

SOLUTION

According to our definition, the area is

$$2\pi \int_1^2 f(x)(1 + [f'(x)]^2)^{1/2}\, dx$$

$$= 2\pi \int_1^2 2x^3(1 + [6x^2]^2)^{1/2}\, dx$$

$$= \frac{\pi}{54} \int_1^2 \frac{3}{2}(1 + 36x^4)^{1/2}(144x^3)\, dx.$$

This integral is easily calculated using the u-substitution $u = 36x^4$, $du = 144x^3\, dx$. With this substitution the limits of integration become 36 and 576; the area is thus equal to

$$\frac{\pi}{54} \int_{36}^{576} \frac{3}{2}(1 + u)^{1/2}\, du = \frac{\pi}{54}(1 + u)^{3/2}\Big|_{36}^{576}$$

$$= \frac{\pi}{54}[(577)^{3/2} - (37)^{3/2}]$$

$$\approx 793.24866.$$

EXAMPLE 8.24

Find the surface area of a right circular cone with base of radius 4 and height 8.

SOLUTION

It is convenient to think of such a cone as the surface obtained by rotating the graph of $f(x) = x/2, 0 \le x \le 8$, about the x-axis (Fig. 8.34). According to our definition, the surface area of the cone is

$$2\pi \int_0^8 \frac{x}{2}[1 + (1/2)^2]^{1/2} \, dx = 2\pi \frac{\sqrt{5}}{4} \int_0^8 x \, dx$$

$$= 16\sqrt{5}\pi.$$

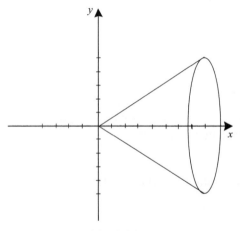

Fig. 8.34

You Try It: The standard formula for the surface area of a cone is

$$S = \pi r \sqrt{h^2 + r^2}.$$

Derive this formula by the method of Example 8.24.

We may also consider the area of a surface obtained by rotating the graph of a function about the y-axis. We do so by using y as the independent variable. Here is an example:

EXAMPLE 8.25

Set up, but do not evaluate, the integral for finding the area of the surface obtained when the graph of $f(x) = x^6, 1 \le x \le 4$, is rotated about the y-axis.

SOLUTION
We think of the curve as the graph of $\phi(y) = y^{1/6}$, $1 \le y \le 4096$. Then the formula for surface area is

$$2\pi \int_{1}^{4096} \phi(y)\left(1 + [\phi'(y)]^2\right)^{1/2} \, dy.$$

Calculating $\phi'(y)$ and substituting, we find that the desired surface area is the value of the integral

$$2\pi \int_{1}^{4096} y^{1/6}\left(1 + \left[(1/6)y^{-5/6}\right]^2\right)^{1/2} \, dy.$$

You Try It: Write the integral that represents the surface area of a hemisphere of radius one and evaluate it.

8.6 Hydrostatic Pressure

If a liquid sits in a tank, then it exerts force on the side of the tank. This force is caused by gravity, and the greater the depth of the liquid then the greater the force. *Pascal's principle* asserts that the force exerted by a body of water depends on depth alone, and is the same in all directions. Thus the force on a point in the side of the tank is defined to be the depth of the liquid at that point times the density of the liquid. Naturally, if we want to design tanks which will not burst their seams, it is important to be able to calculate this force precisely.

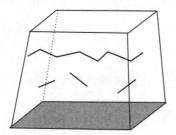

Fig. 8.35

Imagine a tank of liquid having density ρ pounds per cubic foot as shown in Fig. 8.35. We want to calculate the force on one flat side wall of the tank. Thus we will use the independent variable h to denote depth, measured down from the surface of the water, and calculate the force on the wall of the tank between depths $h = a$ and $h = b$ (Fig. 8.36). We partition the interval $[a, b]$:

$$a = h_0 \le h_1 \le h_2 \le \cdots \le h_{k-1} \le h_k = b.$$

Assume that the width of the tank at depth h is $w(h)$. The portion of the wall between $h = h_{j-1}$ and $h = h_j$ is then approximated by a rectangle R_j of length $w(h_j)$ and width $\Delta h = h_j - h_{j-1}$ (Fig. 8.37).

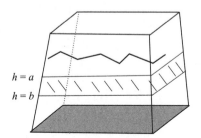

Fig. 8.36

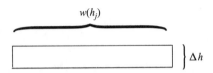

Fig. 8.37

Now we have the following data:

$$\text{Area of Rectangle} = w(h_j) \cdot \Delta h \ \text{square feet}$$
$$\text{Depth of Water} \approx h_j \ \text{feet}$$
$$\text{Density of Liquid} = \rho \ \text{pounds per cubic foot.}$$

It follows that the force exerted on this thin portion of the wall is about

$$P_j = h_j \cdot \rho \cdot w(h_j) \cdot \Delta h.$$

Adding up the force on each R_j gives a total force of

$$\sum_{j=1}^{k} P_j = \sum_{j=1}^{k} h_j \rho \, w(h_j) \Delta h.$$

But this last expression is a Riemann sum for the integral

$$\int_a^b h\rho w(h) \, dh. \tag{$*$}$$

EXAMPLE 8.26

A swimming pool is rectangular in shape, with vertical sides. The bottom of the pool has dimensions 10 feet by 20 feet and the depth of the water is 8 feet. Refer to Fig. 8.38. The pool is full. Calculate the total force on one of the long sides of the pool.

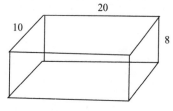

Fig. 8.38

SOLUTION

We let the independent variable h denote depth, measured vertically down from the surface of the water. Since the pool is rectangular with vertical sides, $w(h)$ is constantly equal to 20 (because we are interested in the long side). We use 62.4 pounds per cubic foot for the density of water. According to (∗), the total force on the long side is

$$\int_0^8 h \cdot 62.4 \cdot w(h)\, dh = \int_0^8 h \cdot 62.4 \cdot 20\, dh = 39936 \text{ lbs.}$$

You Try It: A tank full of water is in the shape of a cube of side 10 feet. How much force is exerted against one wall of the tank between the depths of 3 feet and 6 feet?

EXAMPLE 8.27

A tank has vertical cross section in the shape of an inverted isosceles triangle with horizontal base, as shown in Fig. 8.39. Notice that the base of the tank has length 4 feet and the height is 9 feet. The tank is filled with water to a depth of 5 feet. Water has density 62.4 pounds per cubic foot. Calculate the total force on one end of the tank.

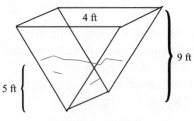

Fig. 8.39

SOLUTION

As shown in Fig. 8.40, at depth h (measured *down* from the surface of the water), the tank has width corresponding to the base of an isosceles triangle similar to the triangle describing the end of the tank. The height of this triangle is $5 - h$. Thus we can solve

$$\frac{w(h)}{5 - h} = \frac{4}{9}.$$

We find that

$$w(h) = \frac{4}{9}(5 - h).$$

According to (∗), the total force on the side is then

$$\int_0^5 h \cdot 62.4 \cdot \frac{4}{9}(5 - h)\,dh \approx 577.778 \text{ lbs.}$$

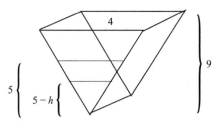

Fig. 8.40

EXAMPLE 8.28

An aquarium tank is filled with a mixture of water and algicide to keep the liquid clear for viewing. The liquid has a density of 50 pounds per cubic foot. For viewing purposes, a window is located in the side of the tank, with center 20 feet below the surface. The window is in the shape of a square of side $4\sqrt{2}$ feet with vertical and horizontal diagonals (see Fig. 8.41). What is the total force on this window?

SOLUTION

As usual, we measure depth downward from the surface with independent variable h. Of course the square window has diagonal 4 feet. Then the range of integration will be $h = 20 - 4 = 16$ to $h = 20 + 4 = 24$. Refer to Fig. 8.42. For h between 16 and 20, we notice that the right triangle in Fig. 8.42 is isosceles and hence has base of length $h - 16$. Therefore

$$w(h) = 2(h - 16) = 2h - 32.$$

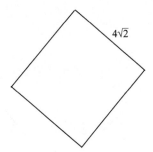

Fig. 8.41

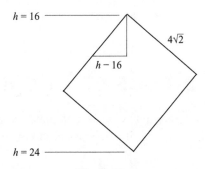

Fig. 8.42

According to our analysis, the total force on the upper half of the window is thus

$$\int_{16}^{20} h \cdot 50 \cdot (2h - 32)\, dh = \frac{44880}{3} \text{ lbs.}$$

For the lower half of the window, we examine the isosceles right triangle in Fig. 8.43. It has base $24 - h$. Therefore, for h ranging from 20 to 24, we have

$$w(h) = 2(24 - h) = 48 - 2h.$$

According to our analysis, the total force on the lower half of the window is

$$\int_{20}^{24} h \cdot 50 \cdot (48 - 2h)\, dh = \frac{51200}{3} \text{ lbs.}$$

The total force on the entire window is thus

$$\frac{44880}{3} + \frac{51200}{3} = \frac{96080}{3} \text{ lbs.}$$

You Try It: A tank of water has flat sides. In one side, with center 4 feet below the surface of the water, is a circular window of radius 1 foot. What is the total force on the window?

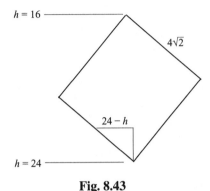

Fig. 8.43

8.7 Numerical Methods of Integration

While there are many integrals that we can calculate explicitly, there are many others that we cannot. For example, it is impossible to evaluate

$$\int e^{-x^2}\, dx. \qquad (*)$$

That is to say, it can be proved mathematically that no closed-form antiderivative can be written down for the function e^{-x^2}. Nevertheless, $(*)$ is one of the most important integrals in all of mathematics, for it is the Gaussian probability distribution integral that plays such an important role in statistics and probability.

Thus we need other methods for getting our hands on the value of an integral. One method would be to return to the original definition, that is to the Riemann sums. If we need to know the value of

$$\int_0^1 e^{-x^2}\, dx$$

then we can approximate this value by a Riemann sum

$$\int_0^1 e^{-x^2}\, dx \approx e^{-(0.25)^2} \cdot 0.25 + e^{-(0.5)^2} \cdot 0.25 + e^{-(0.75)^2} \cdot 0.25 + e^{-1^2} \cdot 0.25.$$

A more accurate approximation could be attained with a finer approximation:

$$\int_0^1 e^{-x^2}\, dx \approx \sum_{j=1}^{10} e^{-(j \cdot 0.1)^2} \cdot 0.1 \qquad (**)$$

or

$$\int_0^1 e^{-x^2}\, dx \approx \sum_{j=1}^{100} e^{-(j\cdot 0.01)^2} \cdot 0.01 \qquad\qquad (\star)$$

The trouble with these "numerical approximations" is that they are calculationally expensive: the degree of accuracy achieved compared to the number of calculations required is not attractive.

Fortunately, there are more accurate and more rapidly converging methods for calculating integrals with numerical techniques. We shall explore some of these in the present section.

It should be noted, and it is nearly obvious to say so, that the techniques of this section require the use of a computer. While the Riemann sum ($\ast\ast$) could be computed by hand with some considerable effort, the Riemann sum ($\star$) is all but infeasible to do by hand. Many times one wishes to approximate an integral by the sum of a thousand terms (if, perhaps, five decimal places of accuracy are needed). In such an instance, use of a high-speed digital computer is virtually mandatory.

8.7.1   THE TRAPEZOID RULE

The method of using Riemann sums to approximate an integral is sometimes called "the method of rectangles." It is adequate, but it does not converge very quickly and it begs more efficient methods. In this subsection we consider the method of approximating by trapezoids.

Let f be a continuous function on an interval $[a, b]$ and consider a partition $\mathcal{P} = \{x_0, x_1, \ldots, x_k\}$ of the interval. As usual, we take $x_0 = a$ and $x_k = b$. We also assume that the partition is uniform.

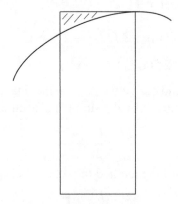

Fig. 8.44

In the method of rectangles we consider a sum of the areas of rectangles. Figure 8.44 shows one rectangle, how it approximates the curve, and what error is made in this particular approximation. The rectangle gives rise to a "triangular" error region (the difference between the true area under the curve and the area of the rectangle). We put quotation marks around the word "triangular" since the region in question is not a true triangle but instead is a sort of curvilinear triangle. If we instead approximate by trapezoids, as in Fig. 8.45 (which, again, shows just one region), then at least visually the errors seem to be much smaller.

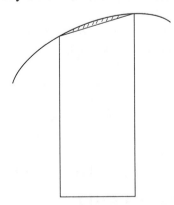

Fig. 8.45

In fact, letting $\triangle x = x_j - x_{j-1}$ as usual, we see that the first trapezoid in the figure has area $[f(x_0) + f(x_1)] \cdot \triangle x/2$. The second has area $[f(x_1) + f(x_2)] \cdot \triangle x/2$, and so forth. In sum, the aggregate of the areas of all the trapezoids is

$$\frac{1}{2} \cdot \{f(x_0) + f(x_1)\} \cdot \triangle x + \frac{1}{2} \cdot \{f(x_1) + f(x_2)\} \cdot \triangle x + \cdots$$
$$+ \frac{1}{2} \cdot \{f(x_{k-1}) + f(x_k)\} \cdot \triangle x$$
$$= \frac{\triangle x}{2} \cdot \{f(x_0) + 2f(x_1) + 2f(x_2)$$
$$+ \cdots + 2f(x_{k-1}) + f(x_k)\}. \tag{†}$$

It is known that, if the second derivative of f on the interval $[a, b]$ does not exceed M then the approximation given by the sum (†) is accurate to within

$$\frac{M \cdot (b-a)^3}{12k^2}.$$

[By contrast, the accuracy of the method of rectangles is generally not better than

$$\frac{N \cdot (b-a)^2}{2k},$$

where N is an upper bound for the *first derivative* of f. We see that the method of trapezoids introduces an extra power of $(b - a)$ in the numerator of the error estimate and, perhaps more importantly, an extra factor of k in the denominator.]

EXAMPLE 8.29

Calculate the integral

$$\int_0^1 e^{-x^2}\, dx$$

to two decimal places of accuracy.

SOLUTION

We first calculate that if $f(x) = e^{-x^2}$ then $f''(x) = (4x^2 - 2)e^{-x^2}$ and therefore $|f''(x)| \leq 2 = M$ for $0 \leq x \leq 1$. In order to control the error, and to have two decimal places of accuracy, we need to have

$$\frac{M \cdot (b-a)^3}{12k^2} < 0.005$$

or

$$\frac{2 \cdot 1^3}{12k^2} < 0.005.$$

Rearranging this inequality gives

$$\frac{100}{3} < k^2.$$

Obviously $k = 6$ will do.

So we will use the partition $\mathcal{P} = \{0, 1/6, 1/3, 1/2, 2/3, 5/6, 1\}$. The corresponding trapezoidal sum is

$$S = \frac{1/6}{2} \cdot \left\{ e^{-0^2} + 2e^{-(1/6)^2} + 2e^{-(1/3)^2} + 2e^{-(1/2)^2} \right.$$
$$\left. + 2e^{-(2/3)^2} + 2e^{-(5/6)^2} + e^{-1^2} \right\}.$$

Some tedious but feasible calculation yields then that

$$S = \frac{1}{12} \cdot \{1 + 2 \cdot .9726 + 2 \cdot .8948 + 2 \cdot .7880$$
$$+ 2 \cdot .6412 + 2 \cdot .4994 + .3679\}$$
$$= \frac{8.9599}{12} = .7451.$$

We may use a computer algebra utility like Mathematica or Maple to calculate the integral *exactly* (to six decimal places) to equal 0.746824. We thus see that the answer we obtained with the Trapezoid Rule is certainly accurate to two decimal places. It is *not* accurate to three decimal places.

It should be noted that Maple and Mathematica both use numerical techniques, like the ones being developed in this section, to calculate integrals. So our calculations merely emulate what these computer algebra utilities do so swiftly and so well.

You Try It: How fine a partition would we have needed to use if we wanted four decimal places of accuracy in the last example? If you have some facility with a computer, use the Trapezoid Rule with *that* partition and confirm that your answer agrees with Mathematica's answer to four decimal places.

EXAMPLE 8.30

Use the Trapezoid Rule with $k = 4$ to estimate

$$\int_0^1 \frac{1}{1+x^2}\, dx.$$

SOLUTION

Of course we could calculate this integral precisely by hand, but the point here is to get some practice with the Trapezoid Rule. We calculate

$$S = \frac{1/4}{2} \cdot \left\{ \frac{1}{1+0^2} + 2\cdot\frac{1}{1+\left(\frac{1}{4}\right)^2} + 2\cdot\frac{1}{1+\left(\frac{2}{4}\right)^2} + 2\cdot\frac{1}{1+\left(\frac{3}{4}\right)^2} + \frac{1}{1+1^2} \right\}.$$

A bit of calculation reveals that

$$S = \frac{1}{8}\cdot\frac{5323}{850} \approx 0.782794\ldots.$$

Now if we take $f(x) = 1/(1+x^2)$ then $f''(x) = (6x^2 - 2)/(1 + x^2)^3$. Thus, on the interval $[0, 1]$, we have that $|f''(x)| \le 4 = M$. Thus the error estimate for the Trapezoid Rule predicts accuracy of

$$\frac{M\cdot(b-a)^3}{12k^2} = \frac{4\cdot 1^3}{12\cdot 4^2} \approx 0.020833\ldots.$$

This suggests accuracy of one decimal place.

Now we know that the true and exact value of the integral is $\arctan 1 \approx 0.78539816\ldots.$ Thus our Trapezoid Rule approximation is good to one, and nearly to two, decimal places—better than predicted.

8.7.2 SIMPSON'S RULE

Simpson's Rule takes our philosophy another step: If rectangles are good, and trapezoids better, then why not approximate by curves? In Simpson's Rule, we approximate by *parabolas*.

We have a continuous function f on the interval $[a, b]$ and we have a partition $\mathcal{P} = \{x_0, x_1, \ldots, x_k\}$ of our partition as usual. It is convenient in this technique to assume that we have an even number of intervals in the partition.

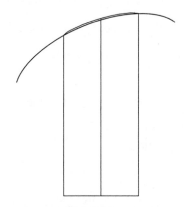

Fig. 8.46

Now each rectangle, over each segment of the partition, is capped off by an arc of a parabola. Figure 8.46 shows just one such rectangle. In fact, for each pair of intervals $[x_{2j-2}, x_{2j-1}]$, $[x_{2j-1}, x_{2j}]$, we consider the unique parabola passing through the endpoints

$$(x_{2j-2}, f(x_{2j-2})), \quad (x_{2j-1}, f(x_{2j-1})), \quad (x_{2j}, f(x_{2j})). \qquad (*)$$

Note that a parabola $y = Ax^2 + Bx + C$ has three undetermined coefficients, so three points as in $(*)$ will determine A, B, C and pin down the parabola.

In fact (pictorially) the difference between the parabola and the graph of f is so small that the error is almost indiscernible. This should therefore give rise to a startling accurate approximation, and it does.

Summing up the areas under all the approximating parabolas (we shall not perform the calculations) gives the following approximation to the integral:

$$\int_a^b f(x)\,dx \approx \frac{\Delta x}{3}\{f(x_0) + 4f(x_1) + 2f(x_2) + 4f(x_3)$$
$$+ 2f(x_4) + \cdots + 2f(x_{k-2}) + 4f(x_{k-1}) + f(x_k)\}.$$

If it is known that the fourth derivative $f^{(iv)}(x)$ satisfies $|f^{(iv)}(x)| \leq M$ on $[a, b]$, then the error resulting from Simpson's method does not exceed

$$\frac{M \cdot (b - a)^5}{180 \cdot k^4}.$$

EXAMPLE 8.31

Use Simpson's Rule to calculate $\int_0^1 e^{-x^2} \, dx$ to two decimal places of accuracy.

SOLUTION

If we set $f(x) = e^{-x^2}$ then it is easy to calculate that

$$f^{(iv)}(x) = e^{-x^2} \cdot [12 - 72x^2 + 32x^4].$$

Thus $|f(x)| \leq 12 = M$.

In order to achieve the desired degree of accuracy, we require that

$$\frac{M \cdot (b-a)^5}{180 \cdot k^4} < 0.005$$

or

$$\frac{12 \cdot 1^5}{180 \cdot k^4} < 0.005.$$

Simple manipulation yields

$$\frac{200}{15} < k^4.$$

This condition is satisfied when $k = 2$.

Thus our job is easy. We take the partition $\mathcal{P} = \{0, 1/2, 1\}$. The sum arising from Simpson's Rule is then

$$S = \frac{1/2}{3} \{f(0) + 4f(1/2) + f(1)\}$$

$$= \frac{1}{6} \{e^{-0^2} + 4 \cdot e^{-(1/2)^2} + e^{-1^2}\}$$

$$= \frac{1}{6} \{1 + 3.1152 + 0.3679\}$$

$$\approx \frac{1}{6} \cdot 4.4831$$

$$\approx 0.7472$$

Comparing with the "exact value" 0.746824 for the integral that we noted in Example 8.29, we find that we have achieved two decimal places of accuracy.

It is interesting to note that if we had chosen a partition with $k = 6$, as we did in Example 8.29, then Simpson's Rule would have guaranteed an accuracy of

$$\frac{M \cdot (b-a)^5}{180 \cdot k^4} = \frac{12 \cdot 1^5}{180 \cdot 6^4} \approx 0.00005144,$$

or nearly four decimal places of accuracy.

EXAMPLE 8.32

Estimate the integral

$$\int_0^1 \frac{1}{1+x^2}\, dx$$

using Simpson's Rule with a partition having four intervals. What degree of accuracy does this represent?

SOLUTION

Of course this example is parallel to Example 8.30, and you should compare the two examples. Our function is $f(x) = 1/(1 + x^2)$ and our partition is $\mathcal{P} = \{0, 1/4, 2/4, 3/4, 1\}$. The sum from Simpson's Rule is

$$S = \frac{1/4}{3} \cdot \{f(0) + 4f(1/4) + 2f(1/2) + 4f(3/4) + f(1)\}$$

$$= \frac{1}{12} \cdot \left\{ \frac{1}{1+0^2} + 4 \cdot \frac{1}{1+(1/4)^2} \right.$$

$$\left. + 2 \cdot \frac{1}{1+(1/2)^2} + 4 \cdot \frac{1}{1+(3/4)^2} + \frac{1}{1+1^2} \right\}$$

$$\approx \frac{1}{12} \cdot \{1 + 3.7647 + 1.6 + 2.56 + 0.5\}$$

$$\approx 0.785392.$$

Comparing with Example 8.30, we see that this answer is accurate to four decimal places. We invite the reader to do the necessary calculation with the Simpson's Rule error to term to confirm that we could have predicted this degree of accuracy.

You Try It: Estimate the integral

$$\int_e^{e^2} \frac{1}{\ln x}\, dx$$

using both the Trapezoid Rule and Simpson's Rule with a partition having six points. Use the error term estimate to state what the accuracy prediction of each of your calculations is. If the software Mathematica or Maple is available to you, check the answers you have obtained against those provided by these computer algebra systems.

Exercises

1. A solid has base the unit circle and vertical slices, parallel to the y-axis, which are half-disks. Calculate the volume of this solid.

2. A solid has base a unit square with center at the origin and vertices on the x- and y-axes. The vertical cross-section of this solid, parallel to the y-axis, is a disk. What is the volume of this solid?

3. Set up the integral to calculate the volume enclosed when the indicated curve over the indicated interval is rotated about the indicated line. Do not evaluate the integral.

   (a) $y = x^2$ $2 \leq x \leq 5$ x-axis
   (b) $y = \sqrt{x}$ $1 \leq x \leq 9$ y-axis
   (c) $y = x^{3/2}$ $0 \leq x \leq 2$ $y = -1$
   (d) $y = x + 3$ $-1 \leq x \leq 2$ $y = 5$
   (e) $y = x^{1/2}$ $4 \leq x \leq 6$ $x = -2$
   (f) $y = \sin x$ $0 \leq x \leq \pi/2$ $y = 0$

4. Set up the integral to evaluate the indicated surface area. Do not evaluate.

   (a) The area of the surface obtained when $y = x^{2/3}$, $0 \leq x \leq 4$, is rotated about the x-axis.
   (b) The area of the surface obtained when $y = x^{1/2}$, $0 \leq x \leq 3$, is rotated about the y-axis.
   (c) The area of the surface obtained when $y = x^2$, $0 \leq x \leq 3$, is rotated about the line $y = -2$.
   (d) The area of the surface obtained when $y = \sin x$, $0 \leq x \leq \pi$, is rotated about the x-axis.
   (e) The area of the surface obtained when $y = x^{1/2}$, $1 \leq x \leq 4$, is rotated about the line $x = -2$.
   (f) The area of the surface obtained when $y = x^3$, $0 \leq x \leq 1$, is rotated about the x-axis.

5. A water tank has a submerged window that is in the shape of a circle of radius 2 feet. The center of this circular window is 8 feet below the surface. Set up, but do not calculate, the integral for the pressure on the lower half of this window—assuming that water weighs 62.4 pounds per cubic foot.

6. A swimming pool is V-shaped. Each end of the pool is an inverted equilateral triangle of side 10 feet. The pool is 25 feet long. The pool is full. Set up, but do not calculate, the integral for the pressure on one end of the pool.

7. A man climbs a ladder with a 100 pound sack of sand that is leaking one pound per minute. If he climbs steadily at the rate of 5 feet per minute, and

if the ladder is 40 feet high, then how much work does he do in climbing the ladder?

8. Because of a prevailing wind, the force that opposes a certain runner is $3x^2 + 4x + 6$ pounds at position x. How much work does this runner perform as he runs from $x = 3$ to $x = 100$ (with distance measured in feet)?

9. Set up, but do not evaluate, the integrals for each of the following arc length problems.

   (a)   The length of the curve $y = \sin x$, $0 \le x \le \pi$
   (b)   The length of the curve $x^2 = y^3$, $1 \le x \le 8$
   (c)   The length of the curve $\cos y = x$, $0 \le y \le \pi/2$
   (d)   The length of the curve $y = x^2$, $1 \le x \le 4$

10. Set up the integral for, but do not calculate, the average value of the given function on the given interval.

   (a) $f(x) = \sin^2 x$ $[2, 5]$
   (b) $g(x) = \tan x$ $[0, \pi/4]$
   (c) $h(x) = \dfrac{x}{x+1}$, $[-2, 2]$
   (d) $f(x) = \dfrac{\sin x}{2 + \cos x}$ $[-\pi, 2\pi]$

11. Write down the sum that will estimate the given integral using the method of rectangles with mesh of size k. You need not actually evaluate the sum.

   (a) $\displaystyle\int_0^4 e^{-x^2}\, dx$ $k = 6$

   (b) $\displaystyle\int_{-2}^2 \sin(e^x)\, dx$ $k = 10$

   (c) $\displaystyle\int_{-2}^0 \cos x^2\, dx$ $k = 5$

   (d) $\displaystyle\int_0^4 \dfrac{e^x}{2 + \sin x}\, dx$ $k = 12$

12. Do each of the problems in Exercise 11 with "method of rectangles" replaced by "trapezoid rule."

13. Do each of the problems in Exercise 11 with "method of rectangles" replaced by "Simpson's Rule."

BIBLIOGRAPHY

[CRC]   Zwillinger et al., *CRC Press Handbook of Tables and Formulas*, 34th ed., CRC Press, Boca Raton, Florida, 1997.

[SCH1] Robert E. Moyer and Frank Ayres, Jr., *Schaum's Outline of Trigonometry*, McGraw-Hill, New York, 1999.

[SCH2] Fred Safier, *Schaum's Outline of Precalculus*, McGraw-Hill, New York, 1997.

[SAH]   S. L. Salas and E. Hille, *Calculus*, John Wiley and Sons, New York, 1982.

SOLUTIONS TO EXERCISES

This book has a great many exercises. For some we provide sketches of solutions and for others we provide just the answers. For some, where there is repetition, we provide no answer. For the sake of mastery, we encourage the student to *write out complete solutions* to all problems.

Chapter 1

1.   (a) $\dfrac{-5}{24}$

    (b) $\dfrac{43219445}{1000000}$

    (c) $\dfrac{-148}{3198}$

    (d) $\dfrac{19800}{34251}$

    (e) $\dfrac{-73162442}{999000}$

    (f) $\dfrac{-108}{705}$

(g) $\dfrac{14}{885}$

(h) $\dfrac{32115422}{9990000}$

2. In Fig. S1.2, set $A = 3.4$, $B = -\pi/2$, $C = 2\pi$, $D = -\sqrt{2}+1$, $E = \sqrt{3}\cdot 4$, $F = 9/2$, $G = -29/10$.

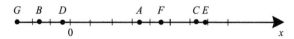

Fig. S1.2

3.

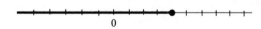

Fig. S1.3(a)

Fig. S1.3(b)

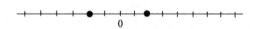

Fig. S1.3(c)

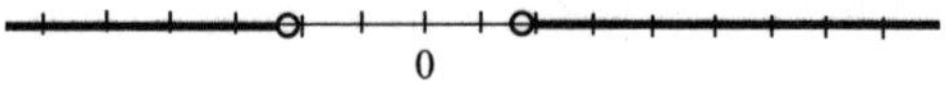

Fig. S1.3(d)

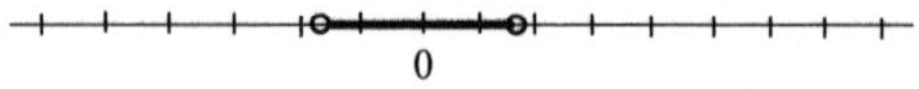

Fig. S1.3(e)

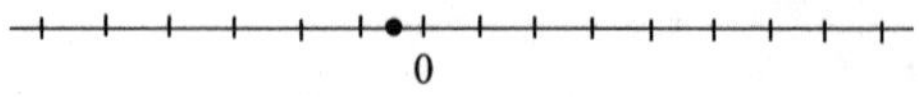

Fig. S1.3(f)

4. Let $A = (2, -4), B = (-6, 3), C = (\pi, \pi^2), D = (-\sqrt{5}, \sqrt{8}), E = (\sqrt{2}\pi, -3),$
$F = (1/3, -19/4).$

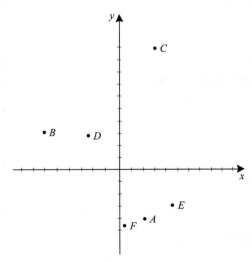

Fig. S1.4

5.

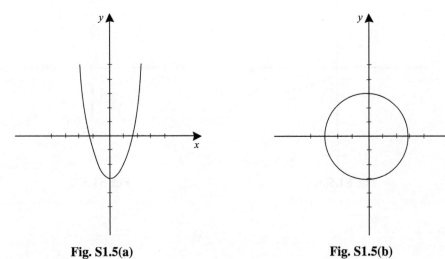

Fig. S1.5(a)   **Fig. S1.5(b)**

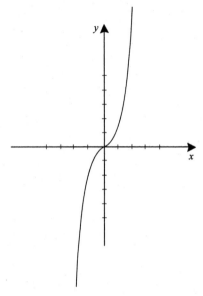

Fig. S1.5(c)

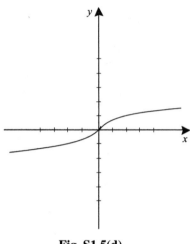

Fig. S1.5(d)

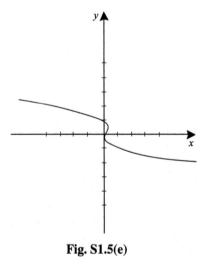

Fig. S1.5(e)

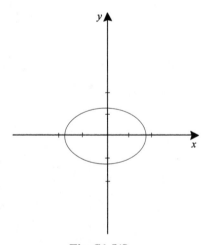

Fig. S1.5(f)

6.

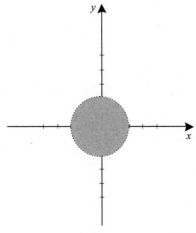

Fig. S1.6(a)

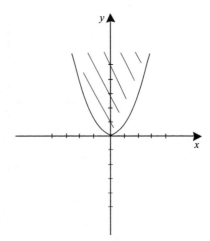

Fig. S1.6(b)

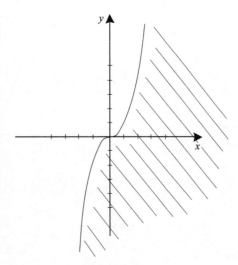

Fig. S1.6(c)

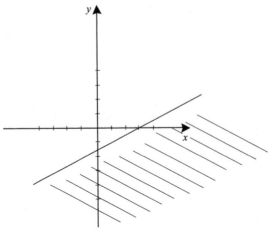

Fig. S1.6(d)

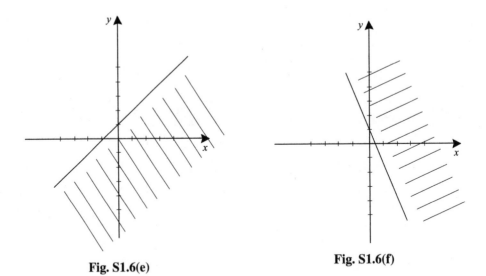

Fig. S1.6(e)

Fig. S1.6(f)

Chapter 1

7. (a) slope $= \dfrac{4 - 6}{2 - (-5)} = \dfrac{-2}{7}$

   (b) Given line has slope $\dfrac{4 - 2}{3 - 1} = 1$ hence requested line has slope -1

   (c) Write $y = -(3/2)x + 3$ hence slope is $-3/2$

   (d) Write $x - 4y = 6x + 6y$ or $y = (-1/2)x$ hence slope is $-1/2$

   (e) slope $= \dfrac{9 - 1}{(-8) - 1} = \dfrac{-8}{9}$

   (f) Write $y = x - 4$ hence slope is 1

8. (a) Slope is $-3/8$ hence line is $y - (-9) = (-3/8) \cdot (x - 4)$

   (b) Slope is 1 hence line is $y - (-8) = 1 \cdot (x - (-4))$

   (c) $y - 6 = (-8)(x - 4)$

   (d) Slope is $\dfrac{3 - 4}{2 - (-6)} = -\dfrac{1}{8}$ hence line is $y - 3 = (-1/8)(x - 2)$

   (e) $y = 6x$

   (f) Slope is -3 hence line is $y - 7 = (-3)(x - (-4))$

9.

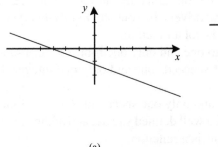

(a)

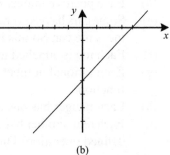

(b)

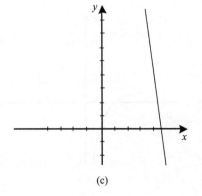

(c)

(d)

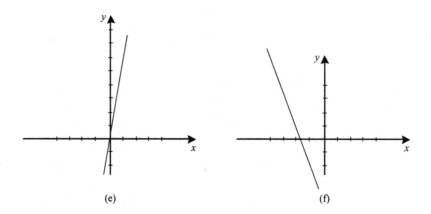

10. (a) Each person has one and only one father. This is a function.
    (b) Some men have more than one dog, others have none. This is not a function.
    (c) Some real numbers have two square roots while others have none. This is not a function.
    (d) Each positive integer has one and only one cube. This is a function.
    (e) Some cars have several drivers. In a one-car family, everyone drives the same car. So this is not a function.
    (f) Each toe is attached to one and only one foot. This is a function.
    (g) Each rational number succeeds one and only one integer. This is a function.
    (h) Each integer has one and only one successor. This is a function.
    (i) Each real number has a well defined square, and adding six is a well defined operation. This is a function.

11.

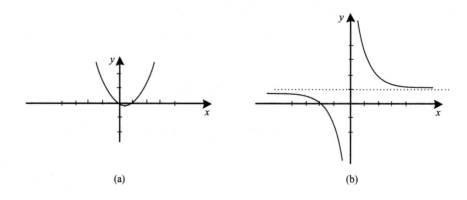

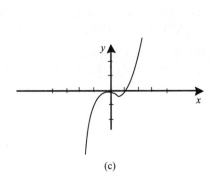

(c)

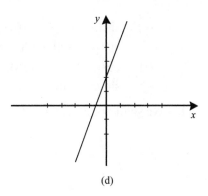

(d)

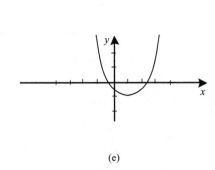

(e)

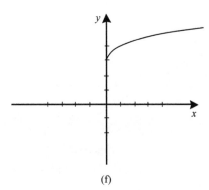

(f)

12.   (a) $\sin(8\pi/3) = \sin(2\pi/3) = \sqrt{3}/2$

     (b) $\tan(-5\pi/6) = [-1/2]/[-\sqrt{3}/2] = 1/\sqrt{3}$

     (c) $\sec(7\pi/4) = 1/\cos(7\pi/4) = \sqrt{2}$

     (d) $\csc(13\pi/4) = \csc(5\pi/4) = 1/\sin(5\pi/4) = 1/[-\sqrt{2}/2] = -\sqrt{2}$

     (e) $\cot(-15\pi/4) = \cot(-7\pi/4) = \cot(\pi/4) = \cos(\pi/4)/\sin(\pi/4) =$
         $[\sqrt{2}/2]/[\sqrt{2}/2] = 1$

     (f) $\cos(-3\pi/4) = -\sqrt{2}/2$

13. We check the first six identities.

     (a) $\cos\pi/3 = 1/2$, $\sin\pi/3 = \sqrt{3}/2$, $\cos^2\pi/3 + \sin^2\pi/3 = [1/2]^2 +$
         $[\sqrt{3}/2]^2 = 1/4 + 3/4 = 1$.

     (b) $\cos\pi/3 = 1/2$, $\sin\pi/3 = \sqrt{3}/2$, $-1 \le 1/2 \le 1$, $-1 \le \sqrt{3}/2 \le 1$.

     (c) $\tan\pi/3 = \sqrt{3}$, $\sec\pi/3 = 2$, $\tan^2\pi/3 + 1 = [\sqrt{3}]^2 + 1 = 3 + 1 =$
         $2^2 = \sec^2\pi/3$.

     (d) $\cot\pi/3 = 1/\sqrt{3}$, $\csc\pi/3 = 2/\sqrt{3}$, $\cot^2\pi/3 + 1 = [1/\sqrt{3}]^2 + 1 =$
         $4/3 = [2/\sqrt{3}]^2 = \csc^2\pi/3$.

(e) $\sin(\pi/3 + (-\pi/6)) = \sin(\pi/6) = 1/2$, $\sin\pi/3\cos(-\pi/6) + \cos\pi/3\sin(-\pi/6) = [\sqrt{3}/2][\sqrt{3}/2] + [1/2][-1/2] = 1/2$.

(f) $\cos(\pi/3 + (-\pi/6)) = \cos(\pi/6) = \sqrt{3}/2$, $\cos\pi/3\cos(-\pi/6) - \sin\pi/3\sin(-\pi/6) = [1/2][\sqrt{3}/2] - [\sqrt{3}/2][-1/2] = \sqrt{3}/2$.

14.   We shall do (a), (c), (e).

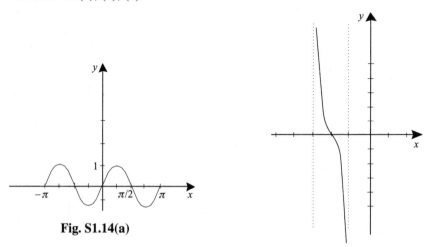

Fig. S1.14(a)

Fig. S1.14(c)

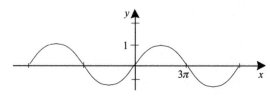

Fig. S1.14(e)

15.   (a) $\theta = (15/2)°$
(b) $\theta = -60°$
(c) $\theta = 405°$
(d) $\theta = (405/4)°$
(e) $\theta = (540/\pi)°$
(f) $\theta = (-900/\pi)°$

16.   (a) $\theta = 13\pi/36$ radians
(b) $\theta = \pi/18$ radians
(c) $\theta = -5\pi/12$ radians

(d) $\theta = -2\pi/3$ radians

(e) $\theta = \pi^2/180$ radians

(f) $\theta = 157\pi/9000$ radians

17. (a) $f \circ g(x) = [(x-1)^2]^2 + 2[(x-1)^2] + 3; g \circ f(x) = ([x^2 + 2x + 3] - 1)^2.$

(b) $f \circ g(x) = \sqrt{\sqrt[3]{x^2 - 2} + 1}; g \circ f(x) = \sqrt[3]{[\sqrt{x+1}]^2 - 2}.$

(c) $f \circ g(x) = \sin([\cos(x^2 - x)]) + 3[\cos(x^2 - x)]^2); g \circ f(x) = \cos([\sin(x + 3x^2)]^2 - [\sin(x + 3x^2)]).$

(d) $f \circ g(x) = e^{\ln(x-5)+2}; g \circ f(x) = \ln(e^{x+2} - 5).$

(e) $f \circ g(x) = \sin([\ln(x^2 - x)]^2 + [\ln(x^2 - x)]); g \circ f(x) = \ln([\sin(x^2 + x)]^2 - [\sin(x^2 + x)]).$

(f) $f \circ g(x) = e^{[e^{-x^2}]^2}; g \circ f(x) = e^{-[e^{x^2}]^2}.$

(g) $f \circ g(x) = [(2x-3)(x+4)] \cdot [(2x-3)(x+4) + 1] \cdot [(2x-3)(x+4) + 2]; g \circ f(x) = (2[(x(x+1)(x+2)] - 3)([(x(x+1)(x+2)] + 4).$

18. (a) f is invertible, with $f^{-1}(t) = (t-5)^{1/3}.$

(b) g is not invertible since $g(0) = g(1) = 0.$

(c) h is invertible, with $h^{-1}(t) = \operatorname{sgn} t \cdot t^2.$

(d) f is invertible, with $f^{-1}(t) = (t-8)^{1/5}.$

(e) g is invertible, with $g^{-1}(t) = -[\ln t]/3.$

(f) h is not invertible, since $\sin \pi/4 = \sin 9\pi/4 = \sqrt{2}/2.$

(g) f is not invertible, since $\tan \pi/4 = \tan 9\pi/4 = 1.$

(h) g is invertible, with $g^{-1}(x) = \operatorname{sgn} x \cdot \sqrt{|x|}.$

19. We will do (a), (c), (e), and (g).

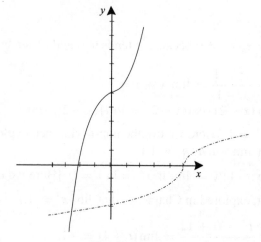

Fig. S1.19(a)

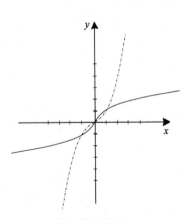

Fig. S1.19(c)

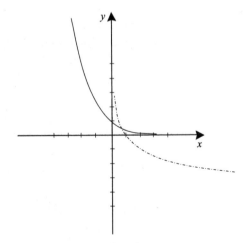

Fig. S1.19(e)

   (g)   Not invertible.

20.   (a)   Invertible, $f^{-1}(t) = \sqrt{t}$.

       (b)   Invertible, $g^{-1}(t) = e^{t}$.

       (c)   Invertible, $h^{-1}(t) = \text{Sin}^{-1}\, t$.

       (d)   Invertible, $f^{-1}(t) = \text{Cos}^{-1}\, t$.

       (e)   Invertible, $g^{-1}(t) = \text{Tan}^{-1}\, t$.

       (f)   Not invertible because $h(-1) = h(1) = 1$.

       (g)   Invertible, $f^{-1}(t) = [3 + \sqrt{9 + 4t}]/2$.

Chapter 2

1.   (a) $\lim\limits_{x \to 0} x \cdot e^{x} = 0$ because x tends to 0 and e^{x} tends to 1.

       (b) $\lim\limits_{x \to 1} \dfrac{x^{2} - 1}{x - 1} = \lim\limits_{x \to 1} x + 1 = 2$.

       (c) $\lim\limits_{x \to 2} (x - 2) \cdot \cot(x - 2) = \lim\limits_{x \to 2} [(x - 2)/\sin(x - 2)] \cdot \cos(x - 2) = 1 \cdot 1 = 1$. [Here we use the non-trivial fact, explored in Chapter 5, that $\lim\limits_{h \to 0} (\sin h / h) = 1$.]

       (d) $\lim\limits_{x \to 0} x \cdot \ln x = \lim\limits_{x \to 0} \ln x^{x} = \ln 1 = 0$. [Here we use the non-trivial fact, explored in Chapter 5, that $\lim\limits_{x \to 0} x^{x} = 1$.]

       (e) $\lim\limits_{t \to 3} \dfrac{t^{2} - 7t + 12}{t - 3} = \lim\limits_{t \to 3} (t - 4) = -1$.

(f) $\lim\limits_{s\to 4} \dfrac{s^2 - 3s - 4}{s - 4} = \lim\limits_{s\to 4}(s + 1) = 5.$

(g) $\lim\limits_{x\to 1} \dfrac{\ln x}{x - 1} = \lim\limits_{x\to 1}\ln[x^{1/(x-1)}] = \lim\limits_{h\to 0}\ln(1 + h)^{1/h} = \ln e = 1.$

[Here we use the non-trivial fact, explored in Chapters 5 and 6, that $\lim\limits_{h\to 0}(1 + h)^{1/h} = e$, where e is Euler's number.]

(h) $\lim\limits_{x\to -3} \dfrac{x^2 - 9}{x + 3} = \lim\limits_{x\to -3} x - 3 = -6.$

2. (a) $\lim\limits_{x\to -1} f(x)$ does not exist, so f is not continuous.

(b) $\lim\limits_{x\to 3} f(x) = 1/2$ and $f(3) = 1/2$ so f is continuous at $c = 3$.

(c) $\lim\limits_{x\to 0} f(x) = 0$. If we define $f(0) = 0$, which is plausible from the graph, then f is continuous at 0.

(d) $\lim\limits_{x\to 0} f(x) = 0$. If we define $f(0) = 0$, which is plausible from the graph, then f is continuous at 0.

(e) $\lim\limits_{x\to 1} f(x) = 1$ and $f(1) = 1$ so f is continuous at $c = 1$.

(f) $\lim\limits_{x\to 1} f(x)$ does not exist so f is not continuous at $c = 1$.

(g) $\lim\limits_{x\to \pi} f(x) = 0$ and $f(0) = 0$ so f is continuous at $c = \pi$.

(h) $\lim\limits_{x\to 2} f(x) = 2 \cdot e^2$ and $f(2) = 2 \cdot e^2$ so f is continuous at $c = 2$.

3. (a) We calculate

$$
\begin{aligned}
f'(2) &= \lim_{h\to 0} \frac{f(2 + h) - f(2)}{h} \\
&= \lim_{h\to 0} \frac{[(2 + h)^2 + 4(2 + h)] - [2^2 + 4 \cdot 2]}{h} \\
&= \lim_{h\to 0} \frac{[4 + 4h + h^2 + 8 + 4h] - [4 + 8]}{h} \\
&= \lim_{h\to 0} \frac{h^2 + 8h}{h} \\
&= \lim_{h\to 0} h + 8 \\
&= 8.
\end{aligned}
$$

The derivative is therefore equal to 8.

(b) We calculate

$$
f'(1) = \lim_{h\to 0} \frac{f(1 + h) - f(h)}{h}
$$

$$= \lim_{h \to 0} \frac{[-1/(1+h)^2] - [-1/1^2]}{h}$$

$$= \lim_{h \to 0} \frac{-1 - [-(1+h)^2]}{h(1+h)^2}$$

$$= \lim_{h \to 0} \frac{2h + h^2}{h + 2h^2 + h^3}$$

$$= \lim_{h \to 0} \frac{2 + h}{1 + 2h + h^2}$$

$$= 2$$

The derivative is therefore equal to 2.

4. (a) $\dfrac{d}{dx} \dfrac{x}{x^2 + 1} = \dfrac{(x^2 + 1) \cdot 1 - x \cdot 2x}{(x^2 + 1)^2} = \dfrac{1 - x^2}{(x^2 + 1)^2}.$

(b) $\dfrac{d}{dx} \sin(x^2) = \left[\dfrac{d}{dx} \sin\right](x^2) \cdot \left(\dfrac{d}{dx} x^2\right) = [\cos(x^2)] \cdot 2x.$

(c) $\dfrac{d}{dt} \tan(t^3 - t^2) = \left[\dfrac{d}{dt} \tan\right](t^3 - t^2) \cdot \dfrac{d}{dt}(t^3 - t^2) = \left[\sec^2(t^3 - t^2)\right] \cdot$
$(3t^2 - 2t).$

(d) $\dfrac{d}{dx}\left(\dfrac{x^2 - 1}{x^2 + 1}\right) = \dfrac{(x^2 + 1) \cdot (2x) - (x^2 - 1) \cdot (2x)}{(x^2 + 1)^2} = \dfrac{4x}{(x^2 + 1)^2}.$

(e) $\dfrac{d}{dx} [x \cdot \ln(\sin x)] = 1 \cdot \ln(\sin x) + x \cdot \dfrac{\cos x}{\sin x} = \ln(\sin x) + \dfrac{x \cdot \cos x}{\sin x}.$

(f) $\dfrac{d}{ds} e^{s(s+2)} = e^{s(s+2)} \cdot [1 \cdot (s + 2) + s \cdot 1] = e^{s(s+2)} \cdot [2s + 2].$

(g) $\dfrac{d}{dx} e^{\sin(x^2)} = e^{\sin(x^2)} \cdot \dfrac{d}{dx} [\sin(x^2)] = e^{\sin(x^2)} \cdot \cos(x^2) \cdot 2x.$

(h) $\left[\ln(e^x + x)\right]' = \dfrac{1}{e^x + x} \cdot (e^x + 1) = \dfrac{e^x + 1}{e^x + x}.$

5. (a) Since the ball is dropped, $v_0 = 0$. The initial height is $h_0 = 100$. Therefore the position of the body at time t is given by

$$p(t) = -16t^2 + 0 \cdot t + 100.$$

The body hits the ground when

$$0 = p(t) = -16t^2 + 100$$

or $t = 2.5$ seconds.

(b) Since the ball has initial velocity 10 feet/second straight down, we know that $v_0 = -10$. The initial height is $h_0 = 100$. Therefore the position of the body at time t is given by

$$p(t) = -16t^2 - 10 \cdot t + 100.$$

The body hits the ground when

$$0 = p(t) = -16t^2 - 10t + 100$$

or $t \approx 2.207$ seconds.

(c) Since the ball has initial velocity 10 feet/second straight up, we know that $v_0 = 10$. The initial height is $h_0 = 100$. Therefore the position of the body at time t is given by

$$p(t) = -16t^2 + 10 \cdot t + 100.$$

The body hits the ground when

$$0 = p(t) = -16t^2 + 10t + 100$$

or $t \approx 2.832$ seconds.

6. (a) $\dfrac{d}{dx} \sin(\ln(\cos x)) = \cos(\ln(\cos x)) \cdot \dfrac{1}{\cos x} \cdot (-\sin x)$

$$= \cos(\ln(\cos x)) \cdot \dfrac{-\sin x}{\cos x}.$$

(b) $\dfrac{d}{dx} e^{\sin(\cos x)} = e^{\sin(\cos x)} \cdot \cos(\cos x) \cdot (-\sin x).$

(c) $\dfrac{d}{dx} \ln(e^{\sin x} + x) = \dfrac{1}{e^{\sin x} + x} \cdot (\cos x + 1).$

(d) $\dfrac{d}{dx} \arcsin(x^2 + \tan x) = \dfrac{1}{\sqrt{1 - [x^2 + \tan x]^2}} \cdot [2x + \sec^2 x].$

(e) $\dfrac{d}{dx} \arccos(\ln x - e^x/5) = \dfrac{-1}{\sqrt{1 - [\ln x - e^x/5]^2}} \cdot \left[\dfrac{1}{x} - \dfrac{e^x}{5} \right].$

(f) $\dfrac{d}{dx} \arctan(x^2 + e^x) = \dfrac{1}{1 + (x^2 + e^x)^2} \cdot [2x + e^x].$

7. Of course $v(t) = p'(t) = 12t - 5$ so $v(4) = 43$ feet/second. The average velocity from $t = 2$ to $t = 8$ is

$$v_{av} = \frac{p(8) - p(2)}{6} = \frac{364 - 34}{6} = 55.$$

The derivative of the velocity function is $(v')'(t) = 12$. This derivative never vanishes, so the extrema of the velocity function on the interval $[5, 10]$ occur at $t = 5$ and $t = 10$. Since $v(5) = 55$ and $v(10) = 115$, we see that the maximum velocity on this time interval is 115 feet per second at $t = 10$.

8. (a) We know that

$$[f^{-1}]'(1) = \frac{1}{f'(0)} = \frac{1}{3}.$$

(b) We know that

$$[f^{-1}]'(1) = \frac{1}{f'(3)} = \frac{1}{8}.$$

(c) We know that

$$[f^{-1}]'(1) = \frac{1}{f'(2)} = \frac{1}{\pi^2}.$$

(d) We know that

$$[f^{-1}]'(1) = \frac{1}{f'(1)} = \frac{1}{40}.$$

Chapter 3

1.

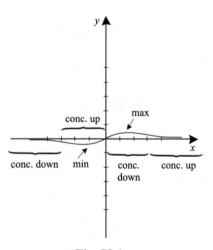

Fig. S3.1

2. Figure S3.2 shows a schematic of the imbedded cylinder. We see that the volume of the imbedded cylinder, as a function of height h, is

$$V(h) = \pi \cdot h \cdot (3 - h/2)^2.$$

Then we solve

$$0 = V'(h) = \pi \cdot \left[9 - 6h + 3h^2/4\right].$$

Fig. S3.2

The roots of this equation are $h = 2, 6$. Of course height 6 gives a trivial cylinder, as does height 0. We find that the solution of our problem is height 2, radius 2.

3. We know that

$$V = \ell \cdot w \cdot h$$

hence

$$\frac{dV}{dt} = \frac{d\ell}{dt} \cdot w \cdot h + \ell \cdot \frac{dw}{dt} \cdot h + \ell \cdot w \cdot \frac{dh}{dt}$$
$$= 1 \cdot 60 \cdot 15 + 100 \cdot (-0.5) \cdot 15 + 100 \cdot 60 \cdot 0.3$$
$$= 900 - 750 + 1800 = 1950 \text{ in/min.}$$

4. We know that $v_0 = -15$. Therefore the position of the body is given by

$$p(t) = -16t^2 - 15t + h_0.$$

Since

$$0 = p(5) = -16 \cdot 5^2 - 15 \cdot 5 + h_0,$$

we find that $h_0 = 475$. The body has initial height 475 feet.

5. We know that

$$V = \frac{1}{3} \cdot \pi r^2 \cdot h.$$

Therefore

$$0 = \frac{d}{dt}V = \frac{1}{3} \cdot \pi \cdot r^2 \cdot \frac{dh}{dt} + \frac{1}{3} \cdot \pi \cdot 2r \cdot \frac{dr}{dt} \cdot h.$$

At the moment of the problem, $dh/dt = 3, r = 5, h = 12/(5\pi)$. Hence

$$0 = \pi \cdot 5^2 \cdot 3^2 + \pi \cdot (2 \cdot 5) \cdot \frac{dr}{dt} \cdot \frac{12}{5\pi}$$

or

$$0 = 225\pi + 24 \cdot \frac{dr}{dt}.$$

We conclude that $dr/dt = -75\pi/8$ microns per minute.

6. Of course

$$10000 = V = \pi \cdot r^2 \cdot h.$$

We conclude that

$$h = \frac{10000}{\pi \cdot r^2}.$$

We wish to minimize

$$A = (\text{area of top}) + (\text{area of sides})$$

$$= \pi \cdot r^2 + 2\pi \cdot r \cdot h = \pi \cdot r^2 + 2\pi \cdot r \cdot \frac{10000}{\pi r^2}.$$

Thus the function to minimize is

$$A(r) = \pi \cdot r^2 + \frac{20000}{r}.$$

Thus

$$0 = A'(r) = 2\pi r - \frac{20000}{r^2}.$$

We find therefore that

$$r^3 = \frac{10000}{\pi}$$

or $r = \sqrt[3]{10000/\pi}$. Since the problem makes sense for $0 < r < \infty$, and since it clearly has no maximum, we conclude that $r = \sqrt[3]{10000/\pi}$, $h = \sqrt[3]{10000/\pi}$.

7. We calculate that $g'(x) = \sin x + x \cos x$ and $g''(x) = 2\cos x - x \sin x$. The roots of these transcendental functions are best estimated with a calculator or computer. Figure S3.7 gives an idea of where the extrema and inflection points are located.

8. We know that $v_0 = -5$ and $h_0 = 400$. Hence

$$p(t) = -16t^2 - 5t + 400.$$

The body hits the ground when

$$0 = p(t) = -16t^2 - 5t + 400.$$

Solving, we find that $t \approx 4.85$ seconds.

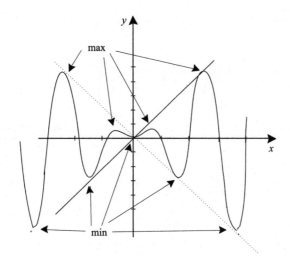

max

min

Fig. S3.7

9. We see that

$$h(x) = \frac{x}{x^2 - 1}$$

$$h'(x) = -\frac{x^2 + 1}{(x^2 - 1)^2}$$

$$h''(x) = \frac{2x(x^2 + 3)}{(x^2 - 1)^3}$$

We see that the function is undefined at ± 1, decreasing everywhere, and has an inflection point only at 0. The sketch is shown in Fig. S3.9.

10. We know that

$$V = \frac{4\pi}{3}r^3.$$

Therefore

$$\frac{dV}{dt} = \frac{4\pi}{3} \cdot 3r^2 \frac{dr}{dt}.$$

Using the values $V = 36\pi, r = 3, dV/dt = -2$, we find that

$$-2 = 4\pi \cdot 3^2 \cdot \frac{dr}{dt}$$

hence

$$\frac{dr}{dt} = -\frac{1}{18\pi} \text{ in. per sec.}$$

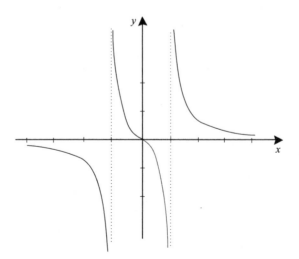

Fig. S3.9

11. The acceleration due to gravity, near the surface of the earth, is about $-32\,\text{ft/sec}^2$ regardless of the mass of the object being dropped. The two stones will strike the ground at the same time.

12. He can drop a rock into the well and time how long it takes the rock to strike the water. Then he can use the equation

$$p(t) = -16t^2 + 0t + h_0$$

to solve for the depth. If the well is *very deep*, then he will have to know the speed of sound and compensate for how long it takes the splash to reach his ears.

13. Refer to Fig. S3.13 to see the geometry of the situation.
Let (x, y) be the point where the rectangle touches the line. Then the area of the rectangle is

$$A = x \cdot y.$$

But of course $3x + 5y = 15$ or $y = 3 - (3/5)x$. Hence

$$A = x \cdot [3 - (3/5)x].$$

We may differentiate and set equal to zero to find that $x = 5/2$ and $y = 3/2$ is the solution to our problem.

14. Let s be a side of the base and let h be the height. The area of the base is s^2 and the same for the top. The area of each side is $s \cdot h$. Thus the cost of the base and top is

$$C_1 = [s^2 + s^2] \cdot 10 \text{ cents}$$

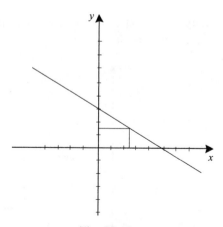

Fig. S3.13

while the cost of the sides is

$$C_2 = 4(s \cdot h) \cdot 20 \text{ cents.}$$

We find that the total cost is

$$C = C_1 + C_2 = 20s^2 + 80sh. \qquad (*)$$

But

$$100 = \text{volume} = s^2 \cdot h$$

hence

$$h = 100/s^2.$$

Substituting this last formula into $(*)$ gives

$$C(s) = 20s^2 + 80s \cdot [100/s^2] = 20s^2 + \frac{8000}{s}.$$

We may calculate that

$$0 = C'(s) = 40s - \frac{8000}{s^2}.$$

Solving for s gives the solution $s \approx 5.8479$ and then $h \approx 2.9241$.

15. We see that

$$f(x) = \frac{x^2 - 1}{x^2 + 1}$$

$$f'(x) = \frac{4x}{(x^2 + 1)^2}$$

$$f''(x) = \frac{-12x^2 + 4}{(x^2 + 1)^3}$$

Thus there are a critical point at $x = 0$ and inflection points at $x = \pm 1/\sqrt{3}$. Figure S3.15 exhibits the complete graph.

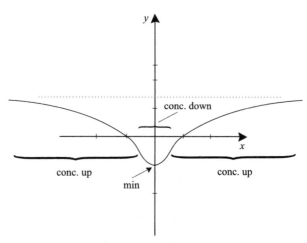

Fig. S3.15

16. We see that the equation for the position of a falling body will now be

$$p(t) = -\frac{20}{2}t^2 + v_0 t + h_0.$$

It is given that $v_0 = 0$ and $h_0 = 100$. Hence

$$p(t) = -10t^2 + 0t + 100.$$

The body hits the surface when

$$0 = p(t) = -10t^2 + 100.$$

This occurs at time $t = \sqrt{10}$.

Chapter 4

1. (a) $F(x) = x^3/3 + \cos x + C$
   (b) $F(x) = e^{3x}/3 + x^5/5 - 2x + C$
   (c) $F(t) = t^3 + [\ln t]^2/2$
   (d) $F(x) = -\ln(\cos x) - \sin x - [\cos 3x]/3 + C$
   (e) $F(x) = [\sin 3x]/3 - [\cos 4x]/4 + x + C$
   (f) $F(x) = e^{\sin x} + C$

2. (a) $\displaystyle \int x \sin x^2\, dx = \frac{-\cos x^2}{2} + C$

(b) $\displaystyle\int \frac{3}{x} \ln x^2 \, dx = \frac{3}{4} \ln^2 x^2 + C$

(c) $\displaystyle\int \sin x \cdot \cos x \, dx = \frac{1}{2} \sin^2 x + C$

(d) $\displaystyle\int \tan x \cdot \ln \cos x \, dx = -\frac{1}{2} \ln^2 \cos x + C$

(e) $\displaystyle\int \sec^2 x \cdot e^{\tan x} \, dx = e^{\tan x} + C$

(f) $\displaystyle\int (2x + 1) \cdot (x^2 + x + 7)^{43} \, dx = \frac{1}{44}(x^2 + x + 7)^{44} + C$

3. (a) We have

$$\int_1^2 x^2 + x \, dx = \lim_{k \to \infty} \sum_{j=1}^k \left[\left(1 + \frac{j}{k}\right)^2 + \left(1 + \frac{j}{k}\right) \right] \cdot \frac{1}{k}$$

$$= \lim_{k \to \infty} \sum_{j=1}^k \left[1 + \frac{2j}{k} + \frac{j^2}{k^2} + 1 + \frac{j}{k} \right] \frac{1}{k}$$

$$= \lim_{k \to \infty} \sum_{j=1}^k \left[\frac{2}{k} + \frac{3j}{k^2} + \frac{j^2}{k^3} \right]$$

$$= \lim_{k \to \infty} \left[k \cdot \frac{2}{k} + \frac{k^2 + k}{2} \cdot \frac{3}{k^2} + \frac{2k^3 + 3k^2 + k}{6} \cdot \frac{1}{k^3} \right]$$

$$= \lim_{k \to \infty} \left[2 + \frac{3}{2} + \frac{3}{2k} + \frac{1}{3} + \frac{1}{2k} + \frac{1}{6k^2} \right]$$

$$= 2 + \frac{3}{2} + \frac{1}{3}$$

$$= \frac{23}{6}.$$

(b) We have

$$\int_{-1}^1 -\frac{x^2}{3} \, dx = \lim_{k \to \infty} \sum_{j=1}^k -\frac{\left(-1 + \frac{2j}{k}\right)^2}{3} \cdot \frac{2}{k}$$

$$= \lim_{k \to \infty} \sum_{j=1}^k \frac{-2}{3k} \left(1 - \frac{4j}{k} + \frac{4j^2}{k^2} \right)$$

$$= \lim_{k \to \infty} \sum_{j=1}^{k} \left(-\frac{2}{3k} + \frac{8j}{3k^2} - \frac{8j^2}{3k^3} \right)$$

$$= \lim_{k \to \infty} k \cdot \frac{-2}{3k} + \frac{k^2 + k}{2} \cdot \frac{8}{3k^2} + \frac{2k^3 + 3k^2 + k}{6} \cdot \frac{-8}{3k^3}$$

$$= -\frac{2}{3} + \frac{4}{3} - \frac{16}{18}$$

$$= -\frac{2}{9}.$$

4.   (a) $\displaystyle\int_{1}^{3} x^2 - 4x^3 + 7\,dx = \left[\frac{x^3}{3} - x^4 + 7x \right]_{1}^{3}$

$$= \left(\frac{3^3}{3} - 3^4 + 7 \cdot 3 \right) - \left(\frac{1^3}{3} - 1^4 + 7 \cdot 1 \right)$$

$$= (9 - 81 + 21) - (1/3 - 1 + 7) = -\frac{172}{3}.$$

(b) $\displaystyle\int_{2}^{6} xe^{x^2} - \sin x \cos x\,dx = \left[\frac{e^{x^2}}{2} - \frac{\sin^2 x}{2} \right]_{2}^{6}$

$$= \left(\frac{e^{36}}{2} - \frac{\sin^2 6}{2} \right) - \left(\frac{e^4}{2} - \frac{\sin^2 2}{2} \right).$$

(c) $\displaystyle\int_{1}^{4} \frac{\ln x}{x} + x \sin x^2\,dx = \left[\frac{\ln^2 x}{2} + \frac{-\cos x^2}{2} \right]_{0}^{4}$

$$= \left(\frac{\ln^2 4}{2} + \frac{-\cos 4^2}{2} \right) - \left(\frac{\ln^2 1}{2} + \frac{-\cos 1^2}{2} \right)$$

$$= \frac{\ln^2 4}{2} - \frac{\cos 16}{2} + \frac{\cos 1}{2}.$$

(d) $\displaystyle\int_{1}^{2} \tan x - x^2 \cos x^3\,dx = \left[-\ln|\cos x| - \frac{\sin x^3}{3} \right]_{1}^{2}$

$$= \left(-\ln|\cos 2| - \frac{\sin 2^3}{3} \right) - \left(-\ln|\cos 1| - \frac{\sin 1^3}{3} \right)$$

$$= -\ln|\cos 2| - \frac{\sin 8}{3} + \ln|\cos 1| + \frac{\sin 1}{3}.$$

(e) $\displaystyle\int_{1}^{e} \frac{\ln x^2}{x}\,dx = \left[\frac{\ln^2 x^2}{4} \right]_{1}^{e} = \frac{\ln^2 e^2}{4} - \frac{\ln^2 1^2}{4} = 1 - 0 = 1.$

(f) $\int_4^8 x^2 \cdot \cos x^3 \sin x^3 \, dx = \left[\dfrac{\sin^2 x^3}{6}\right]_4^8$

$$= \dfrac{\sin^2 8^3}{6} - \dfrac{\sin^2 4^3}{6} = \dfrac{\sin^2 512}{6} - \dfrac{\sin^2 64}{6}.$$

5. (a) Area $= \int_2^5 x^2 + x + 6 \, dx = \left[\dfrac{x^3}{3} + \dfrac{x^2}{2} + 6x\right]_2^5$

$$= \left(\dfrac{5^3}{3} + \dfrac{5^2}{2} + 6 \cdot 5\right) - \left(\dfrac{2^3}{3} + \dfrac{2^2}{2} + 6 \cdot 2\right) = \dfrac{405}{6}.$$

(b) Area $= \int_0^{\pi/4} \sin x \cos x \, dx = \left[\dfrac{\sin^2 x}{2}\right]_0^{\pi/4}$

$$= \dfrac{\sin^2 \pi/4}{2} - \dfrac{\sin^2 0}{2} = \dfrac{1}{4}.$$

(c) Area $= \int_1^2 x e^{x^2} \, dx = \left[\dfrac{e^{x^2}}{2}\right]_1^2 = \dfrac{e^{2^2}}{2} - \dfrac{e^{1^2}}{2} = \dfrac{e^4}{2} - \dfrac{e}{2}.$

(d) Area $= \int_1^e \dfrac{\ln x}{x} \, dx = \left[\dfrac{\ln^2 x}{2}\right]_1^e = \dfrac{\ln^2 e}{2} - \dfrac{\ln^2 1}{2} = \dfrac{1}{2} - \dfrac{0}{2} = \dfrac{1}{2}.$

6.

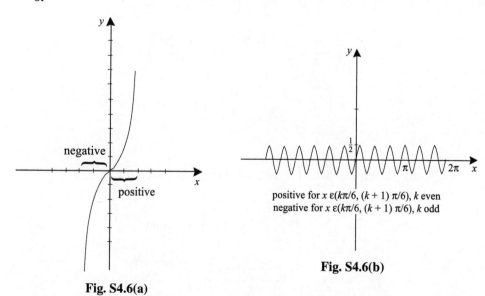

negative

positive

Fig. S4.6(a)

positive for $x \, \varepsilon (k\pi/6, (k+1)\,\pi/6)$, k even
negative for $x \, \varepsilon (k\pi/6, (k+1)\,\pi/6)$, k odd

Fig. S4.6(b)

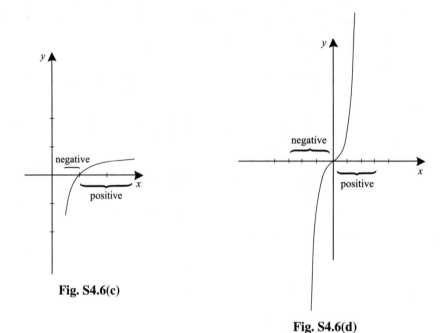

Fig. S4.6(c)

Fig. S4.6(d)

7. (a) Area $= \displaystyle\int_{-2}^{0} -(x^3 + 3x)\, dx + \int_{0}^{2} x^3 + 3x\, dx$

$$= \left[-\left(\frac{x^4}{4} + \frac{3x^2}{2}\right)\right]_{-2}^{0} + \left[\frac{x^4}{4} + \frac{3x^2}{2}\right]_{0}^{2}$$

$$= \left(-\frac{0}{4} - \frac{0}{2} + \frac{(-2)^4}{4} + \frac{3 \cdot (-2)^2}{2}\right)$$

$$+ \left(\frac{2^4}{4} + \frac{3 \cdot 2^2}{2} - \frac{0}{4} - \frac{0}{2}\right) = 20.$$

(b) Area $= \displaystyle\sum_{j=-12}^{11} \int_{j\pi/6}^{(j+1)\pi/6} (-1)^j \sin 3x \cos 3x\, dx$

$$= \sum_{j=-12}^{11} \int_{j\pi/6}^{(j+1)\pi/6} (-1)^j \frac{1}{2} \sin 6x\, dx$$

$$= \sum_{j=-12}^{11} \left[\frac{(-1)^j}{2}\left(-\frac{\cos 6x}{6}\right)\right]_{j\pi/6}^{(j+1)\pi/6}$$

$$= \sum_{j=-12}^{11} \left(\frac{1}{6} - -\frac{1}{6}\right)$$

$$= 8.$$

(c) Area $= \displaystyle\int_{1/2}^{1} -\frac{\ln x}{x}\, dx + \int_{1}^{e} \frac{\ln x}{x}\, dx$

$$= \left[-\frac{\ln^2 x}{2}\right]_{1/2}^{1} + \left[\frac{\ln^2 x}{2}\right]_{1}^{e}$$

$$= \left(-\frac{0}{2} - \left(-\frac{\ln^2(1/2)}{2}\right)\right) + \left(\frac{1^2}{2} - \frac{0^2}{2}\right)$$

$$\approx 0.7404.$$

(d) Area $= \displaystyle\int_{-3}^{0} -x^3 e^{x^4}\, dx + \int_{0}^{3} x^3 e^{x^4}\, dx$

$$= \left[-\frac{e^{x^4}}{4}\right]_{-3}^{0} + \left[\frac{e^{x^4}}{4}\right]_{0}^{3}$$

$$= \left(-\frac{1}{4} - \left(-\frac{e^{81}}{4}\right)\right) + \left(\frac{e^{81}}{4} - \frac{1}{4}\right) = \frac{e^{81}}{2} - \frac{1}{2}.$$

8. (a) Signed Area $= \displaystyle\int_{-2}^{2} x^3 + 3x\, dx = \left[\frac{x^4}{4} + \frac{3x^2}{2}\right]_{-2}^{2}$

$$= \left(\frac{2^4}{4} + \frac{3 \cdot 2^2}{2}\right) - \left(\frac{(-2)^4}{4} + \frac{3 \cdot (-2)^2}{2}\right) = 0.$$

(b) Signed Area $= \displaystyle\int_{-2\pi}^{2\pi} \sin 3x \cos 3x\, dx = \int_{-2\pi}^{2\pi} \frac{1}{2} \sin 6x\, dx$

$$= \left[\frac{1}{2}\left(-\frac{\cos 6x}{6}\right)\right]_{-2\pi}^{2\pi} = -\frac{1}{6} - \left(-\frac{1}{6}\right) = 0.$$

(c) Signed Area $= \displaystyle\int_{1/2}^{e} \frac{\ln x}{x}\, dx = \left[\frac{\ln^2 x}{2}\right]_{1/2}^{e}$

$$= \frac{1^2}{2} - \frac{\ln^2(1/2)}{2}$$

$$\approx 0.2598.$$

(d) Signed Area $= \displaystyle\int_{-3}^{3} x^3 e^{x^4}\, dx = \left[\dfrac{e^{x^4}}{4}\right]_{-3}^{3} = \dfrac{e^{81}}{4} - \dfrac{e^{81}}{4} = 0.$

9. (a) Area $= \displaystyle\int_{-1}^{1} [-3x^2 + 10] - [2x^2 - 4]\, dx = \int_{-1}^{1} -5x^2 + 14\, dx$

$= \left[\dfrac{-5x^3}{3} + 14x\right]_{-1}^{1} = \left(\dfrac{-5}{3} + 14\right) - \left(\dfrac{5}{3} - 14\right) = \dfrac{74}{3}.$

(b) Area $= \displaystyle\int_{0}^{1} x^2 - x^3\, dx = \left[\dfrac{x^3}{3} - \dfrac{x^4}{4}\right]_{0}^{1}$

$= \left(\dfrac{1}{3} - \dfrac{1}{4}\right) - \left(\dfrac{0}{3} - \dfrac{0}{4}\right) = \dfrac{1}{12}.$

(c) Area $= \displaystyle\int_{-3}^{1} [-x^2 + 3] - 2x\, dx = \left[-\dfrac{x^3}{3} + 3x - x^2\right]_{-3}^{1}$

$= \left(-\dfrac{1}{3} + 3 \cdot 1 - 1^2\right) - \left(-\dfrac{-27}{3} + 3 \cdot (-3) - (-3)^2\right)$

$= \dfrac{32}{3}.$

(d) Area $= \displaystyle\int_{1}^{e} x - \ln x\, dx = \left[\dfrac{x^2}{2} - [x \ln x - x]\right]_{1}^{e}$

$= \left(\dfrac{e^2}{2} - [e \cdot 1 - e]\right) - \left(\dfrac{1^2}{2} - [1 \cdot 0 - 1]\right) = \dfrac{e^2 - 3}{2}.$

(e) Area $= \displaystyle\int_{0}^{\pi/4} x - \sin x\, dx = \left[\dfrac{x^2}{2} + \cos x\right]_{0}^{\pi/4}$

$= \left(\dfrac{\pi^2}{32} + \dfrac{\sqrt{2}}{2}\right) - \left(\dfrac{0}{2} + 1\right) = \dfrac{\pi^2}{32} + \dfrac{\sqrt{2}}{2} - 1.$

(f) Area $= \displaystyle\int_{0}^{3} e^x - x\, dx = \left[e^x - \dfrac{x^2}{2}\right]_{0}^{3}$

$= \left(e^3 - \dfrac{9}{2}\right) - \left(e^0 - \dfrac{0}{2}\right) = e^3 - \dfrac{11}{2}.$

10. (a) Area $= \displaystyle\int_{0}^{1} x - x^2\, dx = \left[\dfrac{x^2}{2} - \dfrac{x^3}{3}\right]_{0}^{1} = \left(\dfrac{1}{2} - \dfrac{1}{3}\right) - \left(\dfrac{0}{2} - \dfrac{0}{3}\right)$

$= \dfrac{1}{6}.$

(b) Area $= \int_0^1 \sqrt{x} - x^2 \, dx = \left[\dfrac{x^{3/2}}{3/2} - \dfrac{x^3}{3} \right]_0^1$

$$= \left(\frac{2}{3} - \frac{1}{3} \right) - \left(\frac{0}{3} - \frac{0}{3} \right) = \frac{1}{3}.$$

(c) Area $= \int_{-\sqrt{3}}^{\sqrt{3}} 3x^2 - x^4 \, dx = \left[x^3 - \dfrac{x^5}{5} \right]_{-\sqrt{3}}^{\sqrt{3}} = \left((\sqrt{3})^3 - \dfrac{(\sqrt{3})^5}{5} \right)$

$$- \left((\sqrt{-3})^3 - \frac{(\sqrt{-3})^5}{5} \right) = \frac{12\sqrt{3}}{5}.$$

(d) Area $= \int_{-1}^1 [-2x^2 + 3] - x^4 \, dx = \left[\dfrac{-2x^3}{3} + 3x - \dfrac{x^5}{5} \right]_{-1}^1$

$$= \left(\frac{-2}{3} + 3 - \frac{1}{5} \right) - \left(\frac{2}{3} - 3 - \frac{(-1)}{5} \right) = \frac{64}{15}.$$

(e) Area $= \int_{-2^{1/4}}^{2^{1/4}} [-x^4 + 2] - [x^4 - 2] \, dx = \left[4x - \dfrac{2x^5}{5} \right]_{-2^{1/4}}^{2^{1/4}}$

$$= \left(4 \cdot 2^{1/4} - \frac{2(2^{1/4})^5}{5} \right) - \left(4 \cdot (-2^{1/4}) - \frac{2(-2^{1/4})^5}{5} \right)$$

$$= \frac{32 \cdot 2^{1/4}}{5}.$$

(f) Area $= \int_{-3}^1 [-x^2 + 3] - 2x \, dx = \left[\left(-\dfrac{x^3}{3} + 3x \right) - x^2 \right]_{-3}^1$

$$= \left(-\frac{1}{3} + 3 \cdot 1 - 1^2 \right) - \left(-\frac{-27}{3} + 3 \cdot (-3) - (-3)^2 \right)$$

$$= \frac{32}{3}.$$

Chapter 5

1. (a) $\lim_{x \to 0}(\cos x - 1) = 0$ and $\lim_{x \to 0} x^2 - x^3 = 0$ so l'Hôpital's Rule applies. Thus

$$\lim_{x \to 0} \frac{\cos x - 1}{x^2 - x^3} = \lim_{x \to 0} \frac{-\sin x}{2x - 3x^2}.$$

Now l'Hôpital's Rule applies again to yield

$$= \lim_{x \to 0} \frac{-\cos x}{2 - 6x} = -\frac{1}{2}.$$

(b) $\lim_{x \to 0} e^{2x} - 1 - 2x = 0$ and $\lim_{x \to 0} x^2 + x^4 = 0$ so l'Hôpital's Rule applies. Thus

$$\lim_{x \to 0} \frac{e^{2x} - 1 - 2x}{x^2 + x^4} = \lim_{x \to 0} \frac{2e^{2x} - 2}{2x + 4x^3}.$$

l'Hôpital's Rule applies again to yield

$$= \lim_{x \to 0} \frac{4e^{2x}}{2 + 12x^2} = 2.$$

(c) $\lim_{x \to 0} \cos x \neq 0$, so l'Hôpital's Rule does not apply. In fact the limit does not exist.

(d) $\lim_{x \to 1} [\ln x]^2 = 0$ and $\lim_{x \to 1} (x - 1) = 0$ so l'Hôpital's Rule applies. Thus

$$\lim_{x \to 1} \frac{[\ln x]^2}{(x - 1)} = \lim_{x \to 1} \frac{[2 \ln x]/x}{1} = 0.$$

(e) $\lim_{x \to 2} (x - 2)^3 = 0$ and $\lim_{x \to 2} \sin(x - 2) - (x - 2) = 0$ so l'Hôpital's Rule applies. Thus

$$\lim_{x \to 2} \frac{(x - 2)^3}{\sin(x - 2) - (x - 2)} = \lim_{x \to 2} \frac{3(x - 2)^2}{\cos(x - 2) - 1}.$$

Now l'Hôpital's Rule applies again to yield

$$= \lim_{x \to 2} \frac{6(x - 2)}{-\sin(x - 2)}.$$

We apply l'Hôpital's Rule one last time to obtain

$$= \lim_{x \to 2} \frac{6}{-\cos(x - 2)} = -6.$$

(f) $\lim_{x \to 1} (e^x - 1) = 0$ and $\lim_{x \to 1} (x - 1) = 0$ so l'Hôpital's Rule applies. Thus

$$\lim_{x \to 1} \frac{e^x - 1}{x - 1} = \lim_{x \to 1} \frac{e^x}{1} = e.$$

2. (a) $\lim_{x \to +\infty} x^3 = \lim_{x \to +\infty} (e^x - x^2) = +\infty$ so l'Hôpital's Rule applies. Thus

$$\lim_{x \to +\infty} \frac{x^3}{e^x - x^2} = \lim_{x \to +\infty} \frac{3x^2}{e^x - 2x}.$$

l'Hôpital's Rule applies again to yield

$$= \lim_{x \to +\infty} \frac{6x}{e^x - 2}.$$

l'Hôpital's Rule applies one more time to finally yield

$$\lim_{x \to +\infty} \frac{6}{e^x} = 0.$$

(b) $\lim_{x \to +\infty} \ln x = \lim_{x \to +\infty} x = +\infty$ so l'Hôpital's Rule applies. Thus

$$\lim_{x \to +\infty} \frac{\ln x}{x} = \lim_{x \to +\infty} \frac{1/x}{1} = 0.$$

(c) $\lim_{x \to +\infty} e^{-x} = \lim_{x \to +\infty} \ln[x/(x+1)] = 0$ so l'Hôpital's Rule applies. Thus

$$\lim_{x \to +\infty} \frac{e^{-x}}{\ln[x/(x+1)]} = \lim_{x \to +\infty} \frac{-e^{-x}}{1/x - 1/[x+1]}.$$

It is convenient to rewrite this expression as

$$\lim_{x \to +\infty} \frac{x^2 + x}{-e^x}.$$

Now l'Hôpital's Rule applies once more to yield

$$\lim_{x \to +\infty} \frac{2x + 1}{-e^x}.$$

We apply l'Hôpital's Rule a last time to obtain

$$= \lim_{x \to +\infty} \frac{2}{-e^x} = 0.$$

(d) Since $\lim_{x \to +\infty} \sin x$ does not exist, l'Hôpital's Rule does not apply. In fact the requested limit does not exist.

(e) It is convenient to rewrite this limit as

$$\lim_{x \to -\infty} \frac{x}{e^{-x}}.$$

Since $\lim_{x \to -\infty} x = \lim_{x \to -\infty} e^{-x} = \pm\infty$, l'Hôpital's Rule applies. Thus

$$\lim_{x \to -\infty} \frac{x}{e^{-x}} = \lim_{x \to -\infty} \frac{1}{-e^{-x}} = 0.$$

(f)   Since $\lim_{x \to -\infty} \ln |x| = \lim_{x \to -\infty} e^{-x} = +\infty$, l'Hôpital's Rule applies. Thus

$$\lim_{x \to -\infty} \frac{\ln |x|}{e^{-x}} = \lim_{x \to -\infty} \frac{1/x}{-e^{-x}} = 0.$$

3.   (a)   We write the limit as $\lim_{x \to +\infty} \dfrac{x^3}{e^x}$. Since $\lim_{x \to +\infty} x^3 = \lim_{x \to +\infty} e^x = +\infty$, l'Hôpital's Rule applies. Thus

$$\lim_{x \to +\infty} x^3 e^{-x} = \lim_{x \to +\infty} \frac{x^3}{e^x} = \lim_{x \to +\infty} \frac{3x^2}{e^x}.$$

We apply l'Hôpital's Rule again to obtain

$$= \lim_{x \to +\infty} \frac{6x}{e^x}.$$

Applying l'Hôpital's Rule one last time yields

$$= \lim_{x \to +\infty} \frac{6}{e^x} = 0.$$

(b)   We write the limit as $\lim_{x \to +\infty} \dfrac{\sin(1/x)}{1/x}$. Since $\lim_{x \to +\infty} \sin(1/x) = \lim_{x \to +\infty} 1/x = 0$, l'Hôpital's Rule applies. Hence

$$\lim_{x \to +\infty} x \cdot \sin[1/x] = \lim_{x \to +\infty} \frac{\sin(1/x)}{1/x}$$

$$= \lim_{x \to +\infty} \frac{[\cos(1/x)] \cdot [-1/x^2]}{-1/x^2}$$

$$= \lim_{x \to +\infty} \frac{\cos(1/x)}{1} = 1.$$

(c)   We rewrite the limit as $\lim_{x \to +\infty} \dfrac{\ln[x/(x+1)]}{1/(x+1)}$. Since $\lim_{x \to +\infty} \ln[x/(x+1)] = \lim_{x \to +\infty} 1/(x+1) = 0$, l'Hôpital's Rule applies. Thus

$$\lim_{x \to +\infty} \ln[x/(x+1)] \cdot (x+1) = \lim_{x \to +\infty} \frac{\ln[x/(x+1)]}{1/(x+1)}$$

$$= \lim_{x \to +\infty} \frac{[(x+1)/x] \cdot [1/(x+1)^2]}{-1/(x+1)^2}$$

$$= \lim_{x \to +\infty} \frac{-(x+1)}{x}.$$

Now l'Hôpital's Rule applies again and we obtain

$$= \lim_{x \to +\infty} \frac{-1}{1} = -1.$$

(d) We rewrite the limit as $\lim_{x \to +\infty} \frac{[\ln x]}{e^x}$. Since $\lim_{x \to +\infty} \ln x = \lim_{x \to +\infty} e^x = +\infty$, l'Hôpital's Rule applies. Thus

$$\lim_{x \to +\infty} \ln x \cdot e^{-x} = \lim_{x \to +\infty} \frac{\ln x}{e^x} = \lim_{x \to +\infty} \frac{1/x}{e^x} = 0.$$

(e) We write the limit as $\lim_{x \to -\infty} \frac{x^2}{e^{-2x}}$. Since $\lim_{x \to -\infty} \lim x^2 = \lim_{x \to -\infty} e^{-2x} = 0$, l'Hôpital's Rule applies. Thus

$$\lim_{x \to -\infty} e^{2x} \cdot x^2 = \lim_{x \to -\infty} \frac{x^2}{e^{-2x}} = \lim_{x \to -\infty} \frac{2x}{-2e^{-2x}}.$$

l'Hôpital's Rule applies one more time to yield

$$= \lim_{x \to -\infty} \frac{2}{4e^{-2x}} = 0.$$

(f) We rewrite the limit as $\lim_{x \to 0} \frac{e^{1/x}}{[1/x]}$. Since $\lim_{x \to 0} e^{1/x} = \lim_{x \to 0} 1/x = +\infty$, l'Hôpital's Rule applies. Thus

$$\lim_{x \to 0} x \cdot e^{1/x} = \lim_{x \to 0} \frac{e^{1/x}}{1/x} = \lim_{x \to 0} \frac{e^{1/x} \cdot [-1/x^2]}{-1/x^2} = \lim_{x \to 0} \frac{e^{1/x}}{1} = +\infty.$$

4. We do (a), (b), (c), (d).

(a)
$$\int_0^1 x^{-3/4}\, dx = \lim_{\epsilon \to 0^+} \int_\epsilon^1 x^{-3/4}\, dx = \lim_{\epsilon \to 0^+} \left[\frac{x^{1/4}}{1/4} \right]_\epsilon^1$$

$$= \lim_{\epsilon \to 0^+} \left(\frac{1^{1/4}}{1/4} - \frac{\epsilon^{1/4}}{1/4} \right) = 4.$$

(b)
$$\int_1^3 (x-3)^{-4/3}\, dx = \lim_{\epsilon \to 0^+} \int_1^{3-\epsilon} (x-3)^{-4/3}\, dx$$

$$= \lim_{\epsilon \to 0^+} \left[\frac{(x-3)^{-1/3}}{-1/3} \right]_1^{3-\epsilon} = \lim_{\epsilon \to 0^+} \left(\frac{-\epsilon^{-1/3}}{-1/3} - \frac{-2^{-1/3}}{-1/3} \right). \text{ But}$$
the limit does not exist; so the integral does not converge.

(c)
$$\int_{-2}^2 \frac{1}{(x+1)^{1/3}}\, dx = \lim_{\epsilon \to 0^+} \int_{-2}^{-1-\epsilon} \frac{1}{(x+1)^{1/3}}\, dx$$

$$+ \lim_{\epsilon \to 0^+} \int_{-1+\epsilon}^{2} \frac{1}{(x+1)^{1/3}} \, dx$$

$$= \lim_{\epsilon \to 0^+} \left[\frac{(x+1)^{2/3}}{2/3} \right]_{-2}^{-1-\epsilon} + \lim_{\epsilon \to 0^+} \left[\frac{(x+1)^{2/3}}{2/3} \right]_{-1+\epsilon}^{2}$$

$$= \lim_{\epsilon \to 0^+} \left(\frac{(-\epsilon)^{2/3}}{2/3} - \frac{(-1)^{2/3}}{2/3} \right) + \lim_{\epsilon \to 0^+} \left(\frac{3^{2/3}}{2/3} - \frac{(\epsilon)^{2/3}}{2/3} \right)$$

$$= \frac{3}{2} \cdot \left(3^{2/3} - 1 \right).$$

(d) $\displaystyle \int_{-4}^{6} \frac{x}{(x-1)(x+2)} \, dx = \lim_{\epsilon \to 0^+} \int_{-4}^{-2-\epsilon} \frac{x}{(x-1)(x+2)} \, dx$

$$+ \lim_{\epsilon \to 0^+} \int_{-2+\epsilon}^{0} \frac{x}{(x-1)(x+2)} \, dx + \lim_{\epsilon \to 0^+} \int_{0}^{1-\epsilon} \frac{x}{(x-1)(x+2)} \, dx$$

$$+ \lim_{\epsilon \to 0^+} \int_{1+\epsilon}^{6} \frac{x}{(x-1)(x+2)} \, dx. \text{ Now}$$

$$\frac{x}{(x-1)(x+2)} = \frac{1/3}{x-1} + \frac{2/3}{x+2}.$$

Therefore

$$\int_{-4}^{6} \frac{x}{(x-1)(x+2)} \, dx$$

$$= \lim_{\epsilon \to 0^+} \int_{-4}^{-2-\epsilon} \frac{1/3}{x-1} + \frac{2/3}{x+2} \, dx + \lim_{\epsilon \to 0^+} \int_{-2+\epsilon}^{0} \frac{1/3}{x-1} + \frac{2/3}{x+2} \, dx$$

$$+ \lim_{\epsilon \to 0^+} \int_{0}^{1-\epsilon} \frac{1/3}{x-1} + \frac{2/3}{x+2} \, dx + \lim_{\epsilon \to 0^+} \int_{1+\epsilon}^{6} \frac{1/3}{x-1} + \frac{2/3}{x+2} \, dx$$

$$= \lim_{\epsilon \to 0^+} \left[\frac{1}{3} \ln|x-1| + \frac{2}{3} \ln|x+2| \right]_{-4}^{-2-\epsilon}$$

$$+ \lim_{\epsilon \to 0^+} \left[\frac{1}{3} \ln|x-1| + \frac{2}{3} \ln|x+2| \right]_{-2+\epsilon}^{0}$$

$$+ \lim_{\epsilon \to 0^+} \left[\frac{1}{3} \ln|x-1| + \frac{2}{3} \ln|x+2| \right]_{0}^{1-\epsilon}$$

$$+ \lim_{\epsilon \to 0^+} \left[\frac{1}{3} \ln|x-1| + \frac{2}{3} \ln|x+2| \right]_{1+\epsilon}^{6}.$$

Now this equals

$$\lim_{\epsilon \to 0^+} \left(\frac{1}{3} \cdot \ln|-3-\epsilon| + \frac{2}{3} \ln \epsilon \right) - \left(\frac{1}{3} \cdot \ln 5 + \frac{2}{3} \ln 2 \right) + \text{etc.}$$

The second limit does not exist, so the original integral does not converge.

5. We do (a), (b), (c), (d).

(a) $$\int_1^\infty e^{-3x}\, dx = \lim_{N \to +\infty} e^{-3x}\, dx = \lim_{N \to +\infty} \left[\frac{e^{-3x}}{-3} \right]_1^N$$

$$= \lim_{N \to +\infty} \left(\frac{e^{-3N}}{-3} - \frac{e^{-3}}{-3} \right) = \frac{e^{-3}}{3}.$$

(b) $$\int_2^\infty x^2 e^{-x}\, dx = \lim_{N \to +\infty} \int_2^N x^2 e^{-x}\, dx$$

$$= \lim_{N \to +\infty} \left[-e^{-x} x^2 - 2x e^{-x} - 2 e^{-x} \right]_2^N$$

$$= \lim_{N \to +\infty} \left[(-e^{-N} N^2 - 2N e^{-N} - 2 e^{-N}) \right.$$
$$\left. -(-e^{-2} 2^2 - 2 \cdot 2 \cdot e^{-2} - 2 e^{-2}) \right]$$

$$= e^{-2} 2^2 + 4 e^{-2} + 2 e^{-2}.$$

(c) $$\int_0^\infty x \ln x\, dx = \lim_{\epsilon \to 0^+} \int_\epsilon^1 x \ln x\, dx + \lim_{N \to +\infty} \int_1^N x \ln x\, dx$$

$$= \lim_{\epsilon \to +} [x \ln x - x]_\epsilon^1 + \lim_{N \to +\infty} [x \ln x - x]_1^N$$

$$= \lim_{\epsilon \to 0^+} [(1 \cdot \ln 1 - 1) - (\epsilon \cdot \ln \epsilon - \epsilon)]$$

$$+ \lim_{N \to +\infty} [(N \cdot \ln N - N) - (1 \ln 1 - 1)]$$

$$= \lim_{\epsilon \to 0^+} [-1 + \epsilon] + \lim_{N \to +\infty} [N \ln N - N + 1]$$

$$= \lim_{N \to +\infty} [N \ln N - N]. \text{ This last limit diverges, so}$$

the integral diverges.

(d) $$\int_1^\infty \frac{dx}{1+x^2} = \lim_{N \to +\infty} \int_1^N \frac{dx}{1+x^2} = \lim_{N \to +\infty} [\arctan x]_1^N$$

$$= \lim_{N \to +\infty} (\arctan N - \arctan 1) = \frac{\pi}{2} - \frac{\pi}{4} = \frac{\pi}{4}.$$

Chapter 6

1. (a) $2\ln a - 3\ln b - 4\ln c - \ln d$

   (b) $\dfrac{\ln 2}{\ln 3}$

   (c) $3x + 4\ln z - 3\ln w$

   (d) $2w + \dfrac{1}{2}$

2. We do (a) and (b).

   (a)
   $$3^x \cdot 5^{-x} = 2^x \cdot e^3$$
   $$x\ln 3 - x\ln 5 = x\ln 2 + 3$$
   $$x \cdot [\ln 3 - \ln 5 - \ln 2] = 3$$
   $$x = \frac{3}{\ln 3 - \ln 5 - \ln 2}.$$

   (b)
   $$\frac{3^x}{5^{-x} \cdot 4^{2x}} = 10^x \cdot 10^2$$
   $$x\log_{10} 3 + x\log_{10} 5 - 2x\log_{10} 4 = x + 2$$
   $$x[\log_{10} 3 + \log_{10} 5 - 2\log_{10} 4 - 1] = 2$$
   $$x = \frac{2}{\log_{10} 3 + \log_{10} 5 - 2\log_{10} 4 - 1}.$$

3. (a) $\dfrac{2x \cdot \cos(x^2)}{\sin(x^2)}$

   (b) $\dfrac{2}{x} - \dfrac{1}{x-1}$

   (c) $e^{\sin(e^x)} \cdot \cos(e^x) \cdot e^x$

   (d) $\cos(\ln x) \cdot \dfrac{1}{x}$

4. (a) $-e^{-x}x^3 - e^{-x} \cdot 3x^2 - e^{-x} \cdot 6x - 6e^{-x} + C$

   (b) $\dfrac{x^3}{3}\ln^2 x - \dfrac{2}{9}x^3\ln x + \dfrac{2}{3}\dfrac{x^3}{9} + C$

   (c) $\left[\dfrac{\ln^2 x}{2}\right]_1^e = \dfrac{1}{2}$

   (d) $\left[\ln(e^x + 1)\right]_1^2 = \ln(e^2 + 1) - \ln(e + 1)$

5. We do (a) and (b).

(a) Let $A = x^3 \cdot \dfrac{x^2 + 1}{x^3 - x}$. Then

$$\ln A = 3 \ln x + \ln(x^2 + 1) - \ln(x^3 - x)$$

hence

$$\frac{dA/dx}{A} = \frac{d}{dx} \ln A = \frac{3}{x} + \frac{2x}{x^2 + 1} - \frac{3x^2 - 1}{x^3 - x}.$$

Multiplying through by A gives

$$\frac{dA}{dx} = \left(x^3 \cdot \frac{x^2 + 1}{x^3 - x} \right) \cdot \left[\frac{3}{x} + \frac{2x}{x^2 + 1} - \frac{3x^2 - 1}{x^3 - x} \right].$$

(b) Let $A = \dfrac{\sin x \cdot (x^3 + x)}{x^2(x + 1)}$. Then

$$\ln A = \ln \sin x + \ln(x^3 + x) - \ln x^2 - \ln(x + 1)$$

hence

$$\frac{dA/dx}{A} = \frac{d}{dx} \ln A = \frac{\cos x}{\sin x} + \frac{3x^2 + 1}{x^3 + x} - \frac{2x}{x^2} - \frac{1}{x + 1}.$$

Multiplying through by A gives

$$\frac{dA}{dx} = \left(\frac{\sin x \cdot (x^3 + x)}{x^2(x + 1)} \right) \cdot \left[\frac{\cos x}{\sin x} + \frac{3x^2 + 1}{x^3 + x} - \frac{2x}{x^2} - \frac{1}{x + 1} \right].$$

6. Let $R(t)$ denote the amount of substance present at time t. Let noon on January 10 correspond to $t = 0$ and noon on February 10 correspond to $t = 1$. Then $R(0) = 5$ and $R(1) = 3$. We know that

$$R(t) = P \cdot e^{Kt}.$$

Since

$$5 = R(0) = P \cdot e^{K \cdot 0},$$

we see that $P = 5$. Since

$$3 = R(1) = 5 \cdot e^{K \cdot 1},$$

we find that $K = \ln 3/5$. Thus

$$R(t) = 5 \cdot e^{t \ln(3/5)} = 5 \cdot \left(\frac{3}{5} \right)^t.$$

Taking March 10 to be about $t = 2$, we find that the amount of radioactive material present on March 10 is

$$R(2) = 5 \cdot \left(\frac{3}{5}\right)^2 = \frac{9}{5}.$$

7. Let the amount of bacteria present at time t be

$$B(t) = P \cdot e^{Kt}.$$

Let $t = 0$ be 10:00 a.m. We know that $B(0) = 10000$ and $B(3) = 15000$. Thus

$$10000 = B(0) = P \cdot e^{K \cdot 0}$$

so $P = 10000$. Also

$$15000 = B(3) = 10000 \cdot e^{K \cdot 3}$$

hence

$$K = \frac{1}{3} \cdot \ln(3/2).$$

As a result,

$$B(t) = 10000 \cdot e^{t \cdot [1/3] \ln(3/2)}$$

or

$$B(t) = 10000 \cdot \left(\frac{3}{2}\right)^{t/3}.$$

We find that, at 2:00 p.m., the number of bacteria is

$$B(4) = 10000 \cdot \left(\frac{3}{2}\right)^{4/3}.$$

8. If $M(t)$ is the amount of money in the account at time t then we know that

$$M(t) = 1000 \cdot e^{6t/100}.$$

Here $t = 0$ corresponds to January 1, 2005. Then, on January 1, 2009, the amount of money present is

$$M(4) = 1000 \cdot e^{6 \cdot 4/100} \approx 1271.25.$$

9. (a) $\dfrac{1}{\sqrt{1 - (x \cdot e^x)^2}} \cdot \left[1 \cdot e^x + x \cdot e^x\right]$

(b) $\dfrac{1}{1 + (x/[x+1])^2} \cdot \dfrac{1}{(x+1)^2}$

(c) $\dfrac{1}{1 + [\ln(x^2 + x)]^2} \cdot \dfrac{2x + 1}{x^2 + x}$

(d) $\dfrac{1}{|\tan x| \sqrt{[\tan x]^2 - 1}} \cdot \sec^2 x$

10. (a) $\operatorname{Tan}^{-1} x^2 + C$

(b) $\operatorname{Sin}^{-1} x^3 + C$

(c) $\left[\operatorname{Sin}^{-1}(\sin^2 x) \right]_0^{\pi/2} = \operatorname{Sin}^{-1} 1 - \operatorname{Sin}^{-1} 0 = \dfrac{\pi}{2}.$

(d) $\dfrac{1}{5} \displaystyle\int \dfrac{dx}{1 + [\sqrt{2/5}\,x]^2} = \dfrac{1}{\sqrt{10}} \cdot \operatorname{Tan}^{-1} \left(\dfrac{\sqrt{2}x}{\sqrt{5}} \right) + C$

Chapter 7

1. We do (a), (b), (c), (d).

(a) Let $u = \log^2 x$ and $dv = 1\,dx$. Then

$$\int \log^2 x\,dx = \log^2 x \cdot x - \int x \cdot 2 \log x \cdot \frac{1}{x}\,dx$$

$$= x \log^2 x - 2 \int \log x\,dx.$$

Now let $u = \log x$ and $dv = 1\,dx$. Then

$$\int \log^2 x\,dx = x \log^2 x - 2 \left[\log x \cdot x - \int x \cdot \frac{1}{x}\,dx \right]$$

$$= x \log^2 x - 2x \log x + 2x + C.$$

(b) Let $u = x$ and $dv = e^{3x}\,dx$. Then

$$\int x \cdot e^3 x\,dx = x \cdot \frac{e^{3x}}{3} - \int \frac{e^{3x}}{3} \cdot 1\,dx$$

$$= x \cdot \frac{e^{3x}}{3} - \frac{e^{3x}}{9} + C.$$

(c) Let $u = x^2$ and $dv = \cos x\,dx$. Then

$$\int x^2 \cos x\,dx = x^2 \cdot \sin x - \int \sin x \cdot 2x\,dx.$$

Now let $u = 2x$ and $dv = \sin x\,dx$. Then

$$\int x^2 \cos x\,dx = x^2 \cdot \sin x - \left[2x \cdot (-\cos x) - \int (-\cos x) \cdot 2\,dx \right]$$

$$= x^2 \sin x + 2x \cos x - 2 \sin x + C.$$

(d)   Notice that $\int t \sin 3t \cos 3t \, dt = \frac{1}{2} \int t \sin 6t \, dt$. Now let $u = t$ and $dv = \sin 6t \, dt$. Then

$$\frac{1}{2} \int t \sin 6t \, dt = \frac{1}{2} \left[t \cdot \left(-\frac{1}{6} \cos 6t \right) - \int \left(-\frac{1}{6} \cos 6t \right) \cdot 1 \, dt \right]$$

$$= -\frac{t}{12} \cos 6t + \frac{1}{72} \sin 6t + C.$$

2.   We do (a), (b), (c), (d).

(a) $\dfrac{1}{(x+2)(x-5)} = \dfrac{-1/7}{x+2} + \dfrac{1/7}{x-5}$ hence

$$\int \frac{dx}{(x+2)(x-5)} = \int \frac{-1/7}{x+2} + \int \frac{1/7}{x-5}$$

$$= \frac{-1}{7} \ln|x+2| + \frac{1}{7} \ln|x-5| + C.$$

(b) $\dfrac{1}{(x+1)(x^2+1)} = \dfrac{1/2}{x+1} + \dfrac{-x/2 + 1/2}{x^2+1}$ hence

$$\int \frac{dx}{(x+1)(x^2+1)} = \int \frac{1/2}{x+1} dx + \int \frac{-x/2}{x^2+1} dx + \int \frac{1/2}{x^2+1} dx$$

$$= \frac{1}{2} \ln|x+1| - \frac{1}{4} \ln|x^2+1| + \frac{1}{2} \mathrm{Tan}^{-1} x + C.$$

(c)   Now $x^3 - 2x^2 - 5x + 6 = (x-3)(x+2)(x-1)$. Then

$$\frac{1}{x^3 - 2x^2 - 5x + 6} = \frac{1/10}{x-3} + \frac{1/15}{x+2} + \frac{-1/6}{x-1}.$$

As a result,

$$\int \frac{dx}{x^3 - 2x^2 - 5x + 6} = \int \frac{1/10}{x-3} dx + \int \frac{1/15}{x+2} dx + \int \frac{-1/6}{x-1} dx$$

$$= \frac{1}{10} \ln|x-3| + \frac{1}{15} \ln|x+2| - \frac{1}{6} \ln|x-1| + C.$$

(d)   Now $x^4 - 1 = (x^2 - 1)(x^2 + 1) = (x-1)(x+1)(x^2+1)$. Hence

$$\frac{x}{x^4 - 1} = \frac{1/4}{x-1} + \frac{1/4}{x+1} + \frac{-x/2}{x^2+1}.$$

We conclude that

$$\int \frac{x \, dx}{x^4 + 1} = \frac{1}{4} \ln|x-1| + \frac{1}{4} \ln|x+1| - \frac{1}{4} \ln|x^2+1| + C.$$

3. We do (a), (b), (c), (d).

   (a) Let $u = \sin x$, $du = \cos x \, dx$. Then the integral becomes

   $$\int (1 + u^2)^2 2u \, du = \frac{(1 + u^2)^3}{3} + C.$$

   Resubstituting x, we obtain the final answer

   $$\int (1 + \sin^2 x)^2 2 \sin x \cos x \, dx = \frac{(1 + \sin^2 x)^3}{3} + C.$$

   (b) Let $u = \sqrt{x}$, $du = 1/[2\sqrt{x}] \, dx$. Then the integral becomes

   $$\int 2 \sin u \, du = -2 \cos u + C.$$

   Resubstituting x, we obtain the final answer

   $$\int \frac{\sin \sqrt{x}}{\sqrt{x}} \, dx = -2 \cos \sqrt{x} + C.$$

   (c) Let $u = \ln x$, $du = [1/x] \, dx$. Then the integral becomes

   $$\int \cos u \sin u \, du = \frac{1}{2} \int \sin 2u \, du = -\frac{1}{4} \cos 2u + C.$$

   Resubstituting x, we obtain the final answer

   $$\int \frac{\cos(\ln x) \sin(\ln x)}{x} \, dx = -\frac{1}{4} \cos(2 \ln x) + C.$$

   (d) Let $u = \tan x$, $du = \sec^2 x \, dx$. Then the integral becomes

   $$\int e^u \, du = e^u + C.$$

   Resubstituting x, we obtain the final answer

   $$\int e^{\tan x} \sec^2 x \, dx = e^{\tan x} + C.$$

4. We do (a), (b), (c), (d).

   (a) Let $u = \cos x$, $du = -\sin x \, dx$. Then the integral becomes

   $$-\int u^2 \, du = -\frac{u^3}{3} + C.$$

   Resubstituting x, we obtain the final answer

   $$\int \sin x \cos^2 x \, dx = -\frac{\cos^3 x}{3} + C.$$

(b)   Write

$$\int \sin^3 x \cos^2 x \, dx = \int \sin x (1 - \cos^2 x) \cos^2 x \, dx.$$

Let $u = \cos x$, $du = -\sin x \, dx$. Then the integral becomes

$$-\int (1 - u^2) u^2 \, du = -\frac{u^3}{3} + \frac{u^5}{5} + C.$$

Resubstituting x, we obtain the final answer

$$\int \sin^3 x \cos^2 x \, dx = -\frac{\cos^3 x}{3} + \frac{\cos^5 x}{5} + C.$$

(c)   Let $u = \tan x$, $du = \sec^2 x \, dx$. Then the integral becomes

$$\int u^3 \, du = \frac{u^4}{4} + C.$$

Resubstituting x, we obtain the final answer

$$\int \tan^3 x \sec^2 x \, dx = \frac{\tan^4 x}{4} + C.$$

(d)   Let $u = \sec x$, $du = \sec x \tan x$. Then the integral becomes

$$\int u^2 \, du = \frac{u^3}{3} + C.$$

Resubstituting x, we obtain the final answer

$$\int \tan x \sec^3 x \, dx = \frac{\sec^3 x}{3} + C.$$

5.   We do (a), (b), (c), (d).

(a)   Use integration by parts twice:

$$\int_0^1 e^x \sin x \, dx = \sin x \cdot e^x \Big|_0^1 - \int_0^1 e^x \cos x \, dx$$

$$= [e \cdot \sin 1 - 0] - \left[\cos x e^x \Big|_0^1 - \int_0^1 e^x (-\sin x) \, dx \right]$$

$$= e \cdot \sin 1 - e \cdot \cos 1 + 1 - \int_0^1 e^x \sin x \, dx.$$

We may now solve for the desired integral:

$$\int_0^1 e^x \sin x \, dx = \frac{1}{2} \left[e \cdot \sin 1 - e \cdot \cos 1 \right].$$

(b)　Integrate by parts with $u = \ln x$, $dv = x^2 \, dx$. Thus

$$\int_1^e x^2 \ln x \, dx = \ln x \cdot \left. \frac{x^3}{3} \right|_1^e - \int_1^e \frac{x^3}{3} \cdot \frac{1}{x} \, dx$$

$$= 1 \cdot \frac{e^3}{3} - 0 \cdot \frac{1^3}{3} - \left. \frac{x^3}{6} \right|_1^e$$

$$= \frac{e^3}{3} - \frac{e^3}{9} + \frac{1^3}{9}.$$

(c)　We write

$$\frac{2x+1}{x^2(x+1)} = \frac{1}{x} + \frac{1}{x^2} + \frac{-1}{x+1}.$$

Thus

$$\int_2^4 \frac{(2x+1)\,dx}{x^3 + x^2} = \int_2^4 \frac{1}{x} \, dx + \int_2^4 \frac{1}{x^2} \, dx + \int_2^4 \frac{-1}{x+1} \, dx$$

$$= [\ln 4 - \ln 2] + \left[\frac{-1}{4} - \frac{-1}{2} \right] + [\ln 3 - \ln 5]$$

$$= \ln \frac{6}{5} + \frac{1}{4}.$$

(d)　We write

$$\int_0^\pi \sin^2 x \cos^2 x \, dx = \frac{1}{4} \int_0^\pi \sin^2 2x \, dx$$

$$= \frac{1}{4} \int_0^\pi \frac{1 - \cos 4x}{2} \, dx$$

$$= \frac{1}{8} \left[x - \frac{\sin 4x}{4} \right]_0^\pi$$

$$= \frac{1}{8} \left[(\pi - 0) - (0 - 0) \right]$$

$$= \frac{\pi}{8}.$$

Chapter 8

1. At position x in the base circle, the y-coordinate is $\sqrt{1-x^2}$. Therefore the half-disk slice has radius $\sqrt{1-x^2}$ and area $\pi(1-x^2)/2$. The volume of the solid is then

$$V = \int_{-1}^{1} \frac{\pi(1-x^2)}{2}\,dx$$

$$= \frac{\pi}{2}\left[x - \frac{x^3}{3}\right]_{-1}^{1}$$

$$= \frac{\pi}{2}\left[\left(1 - \frac{1}{3}\right) - \left((-1) - \frac{-1}{3}\right)\right]$$

$$= \frac{2\pi}{3}.$$

2. We calculate the volume of half the solid, and then double the answer. For $0 \leq x \leq 1/\sqrt{2}$, at position x in the base square, the y-coordinate is $1/\sqrt{2} - x$. Thus the disk slice has radius $(1/\sqrt{2} - x)$ and area $\pi(1/\sqrt{2} - x)^2$. Thus the volume of the solid is

$$V = 2\int_{0}^{1/\sqrt{2}} \pi(1/\sqrt{2} - x)^2\,dx$$

$$= \left[-\frac{2\pi}{3}\left(\frac{1}{\sqrt{2}} - x\right)^3\right]_{0}^{1/\sqrt{2}}$$

$$= -\frac{2\pi}{3}\left(0^3 - \left(\frac{1}{\sqrt{2}}\right)^3\right)$$

$$= \frac{\pi}{3\sqrt{2}}.$$

3. (a) $\displaystyle\int_{2}^{5} \pi[x^2]^2\,dx$

   (b) $\displaystyle\int_{1}^{3} \pi[y^2]^2\,dy$

   (c) $\displaystyle\int_{0}^{2} \pi[x^{3/2} + 1]^2\,dx$

   (d) $\displaystyle\int_{-1}^{2} \pi[5 - (x + 3)]^2\,dx$

(e) $\displaystyle\int_2^{\sqrt{6}} \pi[y^2 + 2]^2\, dy$

(f) $\displaystyle\int_0^{\pi/2} \pi[\sin x]^2\, dx$

4. (a) $\displaystyle\int_0^4 2\pi \cdot x^{2/3} \cdot \sqrt{1 + [(2/3)x^{-1/3}]^2}\, dx$

(b) $\displaystyle\int_0^{\sqrt{3}} 2\pi \cdot y^2 \cdot \sqrt{1 + [2y]^2}\, dy$

(c) $\displaystyle\int_0^{\sqrt{3}} 2\pi \cdot [x^2 - 2] \cdot \sqrt{1 + [2x]^2}\, dx$

(d) $\displaystyle\int_0^{\pi} 2\pi \cdot \sin x \cdot \sqrt{1 + [\cos x]^2}\, dx$

(e) $\displaystyle\int_1^2 2\pi \cdot [y^2 + 2] \cdot \sqrt{1 + [2y]^2}\, dy$

(f) $\displaystyle\int_0^1 2\pi \cdot x^3 \cdot \sqrt{1 + [3x^2]^2}\, dx$

5. The depth of points in the window ranges from 6 to 10 feet. At depth x in this range, the window has chord of length $2\sqrt{16x - x^2 - 60}$. Thus the total pressure on the lower half of the window is

$$P = \int_8^{10} 62.4 \cdot x \cdot 2\sqrt{16x - x^2 - 60}\, dx.$$

6. At depth x, the corresponding subtriangle has side-length $2(5 - x/\sqrt{3})$. Therefore the total pressure on one end of the pool is

$$P = \int_0^{5\sqrt{3}} 62.4 \cdot x \cdot 2(5 - x/\sqrt{3})\, dx.$$

7. Let $t = 0$ be the moment when the climb begins. The weight of the sack at time t is then $100 - t$ pounds. Then the work performed during the climb is

$$W = \int_0^8 (100 - t) \cdot 5\, dt.$$

Thus

$$W = \left[500t - \frac{5t^2}{2}\right]_0^8 = \left(4000 - \frac{320}{2}\right) - \left(100 \cdot 0 - \frac{0}{2}\right) = 3840 \text{ ft lbs.}$$

8. The work performed is

$$W = \int_3^{100} [3x^2 + 4x + 6]\, dx$$

$$= \left[x^3 + 2x^2 + 6x \right]_3^{100}$$

$$= (1000000 + 20000 + 600) - (27 + 18 + 18)$$

$$= 1020547 \text{ ft lbs.}$$

9.   (a) $\displaystyle \int_0^\pi \sqrt{1 + [\cos x]^2}\, dx$

      (b) $\displaystyle \int_1^8 \sqrt{1 + [(2/3)x^{-1/3}]^2}\, dx$

      (c) $\displaystyle \int_0^{\pi/2} \sqrt{1 + [-\sin y]^2}\, dy$

      (d) $\displaystyle \int_1^4 \sqrt{1 + [2x]^2}\, dx$

10.   (a) $\displaystyle \frac{1}{3} \int_2^5 \sin^2 x\, dx$

      (b) $\displaystyle \frac{1}{\pi/4} \int_0^{\pi/4} \tan x\, dx$

      (c) $\displaystyle \frac{1}{4} \int_{-2}^2 \frac{x}{x+1}\, dx$

      (d) $\displaystyle \frac{1}{3\pi} \int_{-\pi}^{2\pi} \frac{\sin x}{2 + \cos x}\, dx$

11.   (a) $\displaystyle \sum_{j=1}^6 e^{-(2j/3)^2} \cdot \frac{2}{3}$

      (b) $\displaystyle \sum_{j=1}^{10} \sin(e^{-2+2j/5}) \cdot \frac{2}{5}$

      (c) $\displaystyle \sum_{j=1}^5 \cos(-2 + 2j/5)^2 \cdot \frac{2}{5}$

      (d) $\displaystyle \sum_{j=1}^{12} \frac{e^{j/3}}{2 + \sin(j/3)} \cdot \frac{1}{3}$

12. We do (a) and (b).

(a) $\dfrac{2/3}{2}\Big\{e^{-0^2} + 2\cdot e^{-(2/3)^2} + 2\cdot e^{-(4/3)^2} + 2\cdot e^{-(6/3)^2}$
$\qquad\qquad + 2\cdot e^{-(8/3)^2} + 2\cdot e^{-(10/3)^2} + e^{-(12/3)^2}\Big\}$

(b) $\dfrac{2/5}{2}\Big\{\sin(e^{-10/5}) + 2\cdot\sin(e^{-8/5}) + 2\cdot\sin(e^{-6/5})$
$\qquad\qquad + 2\cdot\sin(e^{-4/5}) + 2\cdot\sin(e^{-2/5}) + 2\cdot\sin(e^{0/5})$
$\qquad\qquad + 2\cdot\sin(e^{2/5}) + 2\cdot\sin(e^{4/5}) + 2\cdot\sin(e^{6/5})$
$\qquad\qquad + 2\cdot\sin(e^{8/5}) + \cdot\sin(e^{10/5})\Big\}$

13. We do (a) and (b)

(a) $\dfrac{2/3}{3}\Big\{e^{-0^2} + 4e^{-(2/3)^2} + 2\cdot e^{-(4/3)^2} + 4\cdot e^{-(6/3)^2}$
$\qquad\qquad + 2\cdot e^{-(8/3)^2} + 4\cdot e^{-(10/3)^2} + e^{-(12/3)^2}\Big\}$

(b) $\dfrac{2/5}{3}\Big\{\sin(e^{-10/5}) + 4\cdot\sin(e^{-8/5}) + 2\cdot\sin(e^{-6/5})$
$\qquad\qquad + 4\cdot\sin(e^{-4/5}) + 2\cdot\sin(e^{-2/5}) + 4\cdot\sin(e^{0/5})$
$\qquad\qquad + 2\cdot\sin(e^{2/5}) + 4\cdot\sin(e^{4/5}) + 2\cdot\sin(e^{6/5})$
$\qquad\qquad + 4\cdot\sin(e^{8/5}) + \sin(e^{10/5})\Big\}$

FINAL EXAM

1. The operations that preserve rational numbers are

    (a)   addition, multiplication, subtraction, division

    (b)   addition and multiplication

    (c)   multiplication and division

    (d)   square roots and logarithm

    (e)   sine and cosine

2. The number $3.157575757\ldots$, expressed as a rational fraction, is

    (a) $\dfrac{3}{2}$

    (b) $\dfrac{2976}{355}$

    (c) $\dfrac{1563}{495}$

    (d) $\dfrac{2}{3}$

    (e) $\dfrac{111}{222}$

3. The number $(\sqrt{3} + \sqrt{2})^2$ is

    (a)   rational

    (b)   irrational

    (c)   transcendental

    (d)   indeterminate

    (e)   the quotient of rational numbers

4. The decimal expansion of $4/7$ is

    (a) $0.213535353\ldots$

    (b) $0.141414114\ldots$

    (c) $0.1357357357\ldots$

    (d) $0.7981818181\ldots$

    (e) $0.571428571428\ldots$

5. The number $\sqrt{3} - \sqrt{2}$

    (a)   lies between 1 and 2

    (b)   is rational

    (c)   is a perfect square

    (d)   lies between -1 and 0

    (e)   lies between 0 and 1

6. The set $\{x : 3 \leq x < 7\}$ is

    (a)   a closed interval

    (b)   an open interval

    (c)   a discrete set

    (d)   a half-open interval

    (e)   a half-line

7. The set $[2, 5) \cap [4, 8]$ is

    (a) $\{x : 4 < x < 8\}$

    (b) $\{t : 4 \leq t < 5\}$

    (c) $\{s : 2 \leq s \leq 4\}$

    (d) $\{w : 2 < w < 8\}$

    (e) $\{u : 4 \leq u \leq 5\}$

8. The set $\mathbb{Q} \cap (-3, 2)$ is

    (a)   infinite

    (b)   finite

    (c)   discrete

    (d)   unbounded

    (e)   arbitrary

9. The set $\{(x, y) : x = y^2\}$ has graph that is

    (a)   a line

(b)   a circle
(c)   a parabola
(d)   a hyperbola
(e)   a directrix

10. The line that passes through the point $(-4, 5)$ and has slope 3 has equation

(a) $x + 3y = 2$
(b) $x - 3y = -4$
(c) $-4x + 5y = 3$
(d) $3x - y = -17$
(e) $3x - 5y = 4$

11. The line $2x + 5y = 10$ has slope

(a)   3
(b)   1
(c)   1/5
(d) $-1/5$
(e) $-2/5$

12. The equation $2x^2 + 2y^2 = 4$ describes

(a)   A circle with center $(0, 0)$ and radius 2
(b)   A circle with center $(0, 0)$ and radius $\sqrt{2}$
(c)   A circle with center $(2, 2)$ and radius 2
(d)   A circle with center $(4, 4)$ and radius 4
(e)   A circle with center $(2, 4)$ and radius 1

13. The equation $x + x^2 + y = 0$ has graph that is

(a)   a circle
(b)   a line
(c)   a parabola
(d)   two crossed lines
(e)   a hyperbola

14. The sine, cosine, and tangent of the angle $5\pi/3$ (measured in radians) are

(a) $1/2, \sqrt{3}/2, \sqrt{3}$
(b) $\sqrt{3}/2, 1/2, 1/\sqrt{3}$
(c) $\sqrt{2}/2, \sqrt{2}/2, 1$
(d) $-\sqrt{3}/2, 1/2, -\sqrt{3}$
(e) $1, 0$, undefined

15. The tangent, cotangent, and secant of the angle $3\pi/4$ (measured in radians) are

   (a) $-\sqrt{3}/2, -1/2, 1$
   (b) $1/\sqrt{2}, 1/\sqrt{2}, -1$
   (c) $\sqrt{2}, -\sqrt{2}, 2$
   (d) $1, -1, \sqrt{3}$
   (e) $-1, -1, -\sqrt{2}$

16. The domain and range of the function $g(x) = \sqrt{1 + 2x}$ are

   (a) $\{x : x \geq -1/2\}$ and $\{x : 0 \leq x < \infty\}$
   (b) $\{x : x \geq 1/2\}$ and $\{x : \sqrt{2} \leq x \leq 2\}$
   (c) $\{x : x \leq -1/2\}$ and $\{y : -2 \leq y < \infty\}$
   (d) $\{s : 1 \leq s \leq 2\}$ and $\{t : 2 \leq t \leq 4\}$
   (e) $\{x : 0 \leq x \leq 2\}$ and $\{x : 1 \leq x \leq 4\}$

17. The graph of the function $y = 1/|x|$ is

   (a) Entirely in the second and third quadrants
   (b) Entirely in the first and fourth quadrants
   (c) Entirely above the x-axis
   (d) Increasing as x moves from left to right
   (e) Decreasing as x moves from left to right

18. The graph of $y = 2x/(1 + x^2)$ includes the points

   (a) $(0, 1), (2, 4), (3, 3)$
   (b) $(1, 1), (2, 2), (4, 4)$
   (c) $(-1, 1), (1, -1), (3, 6)$
   (d) $(1, 1), (2, 4/5), (-2, -4/5)$
   (e) $(0, 0), (-4, 3), (4, 5)$

19. Let $f(x) = x^2 + x$ and $g(x) = x^3 - x$. Then

   (a) $f \circ g(x) = (x^2 + x)^x$ and $g \circ f(x) = (x^2 - x)2x$
   (b) $f \circ g(x) = (x^2 + x)^3 + x$, $g \circ f(x) = (x^3 - x)^2 + x$
   (c) $f \circ g(x) = (x^3 - x)^2 + (x^3 - x)$ and $g \circ f(x) = (x^2 + x)^3 - (x^2 + x)$
   (d) $f \circ g(x) = (x^2 + x) \cdot (x^3 - x)$ and $g \circ f(x) = (x^2 + x)/(x^3 - x)$
   (e) $f \circ g(x) = (x^2 + x) + (x^3 - x)$ and $g \circ f(x) = (x^3 - x)^{x^2 + x}$

20. Let $f(x) = \sqrt[3]{x + 1}$. Then

   (a) $f^{-1}(x) = x^3 - 1$
   (b) $f^{-1}(x) = \sqrt[3]{x} - 1$

(c) $f^{-1}(x) = x^3 - x$

(d) $f^{-1}(x) = x/(x+1)$

(e) $f^{-1}(x) = x^3 - 1$

21. The expression $\ln \dfrac{a^3 \cdot b^{-2}}{c^4/d^{-3}}$ simplifies to

(a) $3 \ln a - 2 \ln b - 4 \ln c + 3 \ln d$

(b) $3 \ln a + 2 \ln b + 4 \ln c - 3 \ln d$

(c) $4 \ln a - 3 \ln b + 2 \ln c - 4 \ln d$

(d) $3 \ln a - 4 \ln b + 3 \ln c - 2 \ln d$

(e) $4 \ln a - 2 \ln b + 2 \ln c + 2 \ln d$

22. The expression $e^{\ln a^2 - \ln b^3}$ simplifies to

(a) $2a \cdot 3b$

(b) $\dfrac{2a}{3b}$

(c) $a^2 \cdot b^3$

(d) $\dfrac{a^2}{b^3}$

(e) $6a^2 b^3$

23. The function $f(x) = \begin{cases} x^2 & \text{if } x < 1 \\ x & \text{if } x \geq 1 \end{cases}$ has limits

(a) 2 at $c = 1$ and -1 at $c = 0$

(b) 1 at $c = 1$ and 4 at $c = -2$

(c) 0 at $c = 0$ and 3 at $c = 5$

(d) -3 at $c = -3$ and 2 at $c = 1$

(e) 1 at $c = 0$ and 2 at $c = 2$

24. The function $f(x) = \dfrac{x}{x^2 - 1}$ has limits

(a) 3 at $c = 1$ and 2 at $c = -1$

(b) ∞ at $c = 1$ and 0 at $c = -1$

(c) 0 at $c = 0$ and nonexistent at $c = \pm 1$

(d) 2 at $c = -2$ and -2 at $c = 2$

(e) $-\infty$ at $c = 1$ and $+\infty$ at $c = -1$

25. The function $f(x) = \begin{cases} x^3 & \text{if } x < 2 \\ \sqrt{x} & \text{if } x \geq 2 \end{cases}$ is continuous at

(a) $x = 2$ and $x = 3$

(b) $x = 2$ and $x = -2$
(c) $x = -2$ and $x = 4$
(d) $x = 0$ and $x = 2$
(e) $x = 2$ and $x = 2.1$

26.   The limit expression that represents the derivative of $f(x) = x^2 + x$ at $c = 3$ is

(a) $\lim\limits_{h \to 0} \dfrac{[(3+h)^2 + (3+h)] - [3^2 + 3]}{h}$

(b) $\lim\limits_{h \to 0} \dfrac{[(3+2h)^2 + (3+h)] - [3^2 + 3]}{h}$

(c) $\lim\limits_{h \to 0} \dfrac{[(3+h)^2 + (3+h)] - [3^2 + 3]}{h^2}$

(d) $\lim\limits_{h \to 0} \dfrac{[(3+h)^2 + (3+2h)] - [3^2 + 3]}{h}$

(e) $\lim\limits_{h \to 0} \dfrac{[(3+h)^2 + (3+h)] - [3^2 + 4]}{h}$

27.   If $f(x) = \dfrac{x - 3}{x^2 + x}$ then

(a) $f'(x) = \dfrac{1}{2x + 1}$

(b) $f'(x) = \dfrac{x^2 - x}{x - 3}$

(c) $f'(x) = (x - 3) \cdot (x^2 + x)$

(d) $f'(x) = \dfrac{-x^2 + 6x + 3}{(x^2 + x)^2}$

(e) $f'(x) = \dfrac{x^2 + 6x - 3}{x^2 + x}$

28.   If $g(x) = x \cdot \sin x^2$ then

(a) $f'(x) = \sin x^2$
(b) $f'(x) = 2x^2 \sin x^2$
(c) $f'(x) = x^3 \sin x^2$
(d) $f'(x) = x \cos x^2$
(e) $f'(x) = \sin x^2 + 2x^2 \cos x^2$

29.   If $h(x) = \ln[x \cos x]$ then

(a) $h'(x) = \dfrac{1}{x \cos x}$

(b) $h'(x) = \dfrac{x \sin x}{x \cos x}$

(c) $h'(x) = \dfrac{\cos x - x \sin x}{x \cos x}$

(d) $h'(x) = x \cdot \sin x \cdot \ln x$

(e) $h'(x) = \dfrac{x \cos x}{\sin x}$

30. If $g(x) = [x^3 + 4x]^{53}$ then

   (a) $g'(x) = 53 \cdot [x^3 + 4x]^{52}$
   (b) $g'(x) = 53 \cdot [x^3 + 4x]^{52} \cdot (3x^2 + 4)$
   (c) $g'(x) = (3x^2 + 4) \cdot 53x^3$
   (d) $g'(x) = x^3 \cdot 4x$
   (e) $g'(x) = \dfrac{x^3 + 4x}{2x^2 + 1}$

31. Suppose that a steel ball is dropped from the top of a tall building. It takes the ball 7 seconds to hit the ground. How tall is the building?

   (a) 824 feet
   (b) 720 feet
   (c) 550 feet
   (d) 652 feet
   (e) 784 feet

32. The position in feet of a moving vehicle is given by $8t^2 - 6t + 142$. What is the acceleration of the vehicle at time $t = 5$ seconds?

   (a) 12 ft/sec^2
   (b) 8 ft/sec^2
   (c) -10 ft/sec^2
   (d) 20 ft/sec^2
   (e) 16 ft/sec^2

33. Let $f(x) = x^3 - 5x^2 + 3x - 6$. Then the graph of f is

   (a) concave up on $(-3, \infty)$ and concave down on $(-\infty, -3)$
   (b) concave up on $(5, \infty)$ and concave down on $(-\infty, 5)$
   (c) concave up on $(5/3, \infty)$ and concave down on $(-\infty, 5/3)$
   (d) concave up on $(3/5, \infty)$ and concave down on $(-\infty, 3/5)$
   (e) concave up on $(-\infty, 5/3)$ and concave down on $(5/3, \infty)$

34. Let $g(x) = x^3 + \frac{7}{2}x^2 - 10x + 2$. Then the graph of f is

   (a) increasing on $(-\infty, -10/3)$ and decreasing on $(-10/3, \infty)$
   (b) increasing on $(-\infty, 1)$ and $(10, \infty)$ and decreasing on $(1, 10)$
   (c) increasing on $(-\infty, -10/3)$ and $(1, \infty)$ and decreasing on $(-10/3, 1)$
   (d) increasing on $(-10/3, \infty)$ and decreasing on $(-\infty, -10/3)$
   (e) increasing on $(-\infty, -10)$ and $(1, \infty)$ and decreasing on $(-10, 1)$

35. Find all local maxima and minima of the function $h(x) = -(4/3)x^3 + 5x^2 - 4x + 8$.

   (a) local minimum at $x = 1/2$, local maximum at $x = 2$
   (b) local minimum at $x = 1/2$, local maximum at $x = 1$
   (c) local minimum at $x = -1$, local maximum at $x = 2$
   (d) local minimum at $x = 1$, local maximum at $x = 3$
   (e) local minimum at $x = 1/2$, local maximum at $x = 1/4$

36. Find all local and global maxima and minima of the function $h(x) = x + 2\sin x$ on the interval $[0, 2\pi]$.

   (a) local minimum at $4\pi/3$, local maximum at $2\pi/3$, global minimum at 0, global maximum at 2π
   (b) local minimum at $2\pi/3$, local maximum at $4\pi/3$, global minimum at 0, global maximum at 2π
   (c) local minimum at 2π, local maximum at 0, global minimum at $4\pi/3$, global maximum at $2\pi/3$
   (d) local minimum at $2\pi/3$, local maximum at 2π, global minimum at $4\pi/3$, global maximum at 0
   (e) local minimum at 0, local maximum at $2\pi/3$, global minimum at $4\pi/3$, global maximum at 2π

37. Find all local and global maxima and minima of the function $f(x) = x^3 + x^2 - x + 1$.

   (a) local minimum at -1, local maximum at $1/3$
   (b) local minimum at 1, local maximum at $-1/3$
   (c) local minimum at 1, local maximum at -1
   (d) local minimum at $1/3$, local maximum at -1
   (e) local minimum at -1, local maximum at 1

38. A cylindrical tank is to be constructed to hold 100 cubic feet of liquid. The sides of the tank will be constructed of material costing $1 per

square foot, and the circular top and bottom of material costing $2 per square foot. What dimensions will result in the most economical tank?

(a) height $= 4 \cdot \sqrt[3]{\pi/25}$, radius $= \sqrt[3]{\pi/25}$

(b) height $= \sqrt[3]{25/\pi}$, radius $= 4 \cdot \sqrt[3]{25/\pi}$

(c) height $= 5^{1/3}$, radius $= \pi^{1/3}$

(d) height $= 4$, radius $= 1$

(e) height $= 4 \cdot \sqrt[3]{25/\pi}$, radius $= \sqrt[3]{25/\pi}$

39. A pigpen is to be made in the shape of a rectangle. It is to hold 100 square feet. The fence for the north and south sides costs $8 per running foot, and the fence for the east and west sides costs $10 per running foot. What shape will result in the most economical pen?

(a) north/south $= 4\sqrt{5}$, east/west $= 5\sqrt{5}$

(b) north/south $= 5\sqrt{5}$, east/west $= 4\sqrt{5}$

(c) north/south $= 4\sqrt{4}$, east/west $= 5\sqrt{4}$

(d) north/south $= 5\sqrt{4}$, east/west $= 4\sqrt{4}$

(e) north/south $= \sqrt{5}$, east/west $= \sqrt{4}$

40. A spherical balloon is losing air at the rate of 2 cubic inches per minute. When the radius is 12 inches, at what rate is the radius changing?

(a) $1/[288\pi]$ in./min

(b) -1 in./min

(c) -2 in./min

(d) $-1/[144\pi]$ in./min

(e) $-1/[288\pi]$ in./min

41. Under heat, a rectangular plate is changing shape. The length is increasing by 0.5 inches per minute and the width is decreasing by 1.5 inches per minute. How is the area changing when $\ell = 10$ and $w = 5$?

(a) The area is decreasing by 9.5 inches per minute.

(b) The area is increasing by 13.5 inches per minute.

(c) The area is decreasing by 10.5 inches per minute.

(d) The area is increasing by 8.5 inches per minute.

(e) The area is decreasing by 12.5 inches per minute.

42. An arrow is shot straight up into the air with initial velocity 50 ft/sec. After how long will it hit the ground?

    (a)   12 seconds
    (b)   25/8 seconds
    (c)   25/4 seconds
    (d)   8/25 seconds
    (e)   8 seconds

43. The set of antiderivates of $x^2 - \cos x + 4x$ is

    (a) $\dfrac{x^3}{3} - \sin x + 2x^2 + C$

    (b) $x^3 + \cos x + x^2 + C$

    (c) $\dfrac{x^3}{4} - \sin x + x^2 + C$

    (d) $x^2 + x + 1 + C$

    (e) $\dfrac{x^3}{2} - \cos x - 2x^2 + C$

44. The indefinite integral $\displaystyle\int \dfrac{\ln x}{x} + x \, dx$ equals

    (a) $\ln x^2 + \ln^2 x + C$

    (b) $\dfrac{\ln^2 x}{2} + \dfrac{x^2}{2} + C$

    (c) $\ln x + \dfrac{1}{\ln x} + C$

    (d) $x \cdot \ln x + C$

    (e) $x^2 \cdot \ln x^2 + C$

45. The indefinite integral $\displaystyle\int 2x \cos x^2 \, dx$ equals

    (a) $[\cos x]^2 + C$
    (b) $\cos x^2 + C$
    (c) $\sin x^2 + C$
    (d) $[\sin x]^2 + C$
    (e) $\sin x \cdot \cos x$

46. The area between the curve $y = -x^4 + 3x^2 + 4$ and the x-axis is

    (a)   20
    (b)   18

(c) 10

(d) $\dfrac{96}{5}$

(e) $\dfrac{79}{5}$

47. The area between the curve $y = \sin 2x + 1/2$ and the x-axis for $0 \le x \le 2\pi$ is

   (a) $2\sqrt{3} - \dfrac{\pi}{3}$

   (b) $-2\sqrt{3} + \dfrac{\pi}{3}$

   (c) $2\sqrt{3} + \dfrac{\pi}{3}$

   (d) $\sqrt{3} + \pi$

   (e) $\sqrt{3} - \pi$

48. The area between the curve $y = x^3 - 9x^2 + 26x - 24$ and the x-axis is

   (a) 3/4

   (b) 2/5

   (c) 2/3

   (d) 1/2

   (e) 1/3

49. The area between the curves $y = x^2 + x + 1$ and $y = -x^2 - x + 13$ is

   (a) $\dfrac{122}{3}$

   (b) $\dfrac{125}{3}$

   (c) $\dfrac{111}{3}$

   (d) $\dfrac{119}{3}$

   (e) $\dfrac{97}{3}$

50. The area between the curves $y = x^2 - x$ and $y = 2x + 4$ is

   (a) $\dfrac{117}{6}$

   (b) $\dfrac{111}{6}$

(c) $\dfrac{125}{6}$

(d) $\dfrac{119}{6}$

(e) $\dfrac{121}{12}$

51. If $\int_1^5 f(x)\,dx = 7$ and $\int_3^5 f(x)\,dx = 2$ then $\int_1^3 f(x)\,dx =$

(a) 4
(b) 5
(c) 6
(d) 7
(e) 3

52. If $F(x) = \int_x^{x^2} \ln t\,dt$ then $F'(x) =$

(a) $(4x - 1) \cdot \ln x$
(b) $x^2 - x$
(c) $\ln x^2 - \ln x$
(d) $\ln(x^2 - x)$
(e) $\dfrac{1}{x^2} - \dfrac{1}{x}$

53. Using l'Hôpital's Rule, the limit $\lim\limits_{x\to 0} \dfrac{\cos 2x - 1}{x^2}$ equals

(a) 1
(b) 0
(c) -4
(d) -2
(e) 4

54. Using l'Hôpital's Rule, the limit $\lim\limits_{x\to +\infty} \dfrac{x^2}{e^{3x}}$ equals

(a) -1
(b) 1
(c) $-\infty$
(d) 0
(e) $+\infty$

55. The limit $\lim\limits_{x\to 0} x^{\sqrt{x}}$ equals

(a) 1

(b) −1

(c) 0

(d) +∞

(e) 2

56. The limit $\lim\limits_{x \to +\infty} \sqrt[3]{x+1} - \sqrt[3]{x}$ equals

(a) 2

(b) 1

(c) 0

(d) −2

(e) −1

57. The improper integral $\int_1^4 \dfrac{1}{\sqrt{x-1}}\,dx$ equals

(a) $\sqrt{3} - 1$

(b) $2(\sqrt{3} - 1)$

(c) $2(\sqrt{3} + 1)$

(d) $\sqrt{3} + 1$

(e) $\sqrt{3}$

58. The improper integral $\int_1^{\infty} \dfrac{x}{1+x^4}\,dx$ equals

(a) $\dfrac{\pi}{3}$

(b) $\dfrac{\pi}{2}$

(c) $\dfrac{\pi}{8}$

(d) $\dfrac{2\pi}{3}$

(e) $\dfrac{3\pi}{4}$

59. The area under the curve $y = x^{-4}$, above the x-axis, and from 3 to $+\infty$, is

(a) $\dfrac{2}{79}$

(b) $\dfrac{1}{79}$

(c) $\dfrac{2}{97}$

(d) $\dfrac{2}{81}$

(e) $\dfrac{1}{81}$

60. The value of $\log_2(1/16) - \log_3(1/27)$ is

     (a)   2
     (b)   3
     (c)   4
     (d)   1
     (e) -1

61. The value of $\dfrac{\log_2 27}{\log_2 3}$ is

     (a) -1
     (b)   2
     (c)   0
     (d)   3
     (e) -3

62. The graph of $y = \ln[1/x^2]$, $x \neq 0$, is

     (a)   concave up for all $x \neq 0$
     (b)   concave down for all $x \neq 0$
     (c)   concave up for $x < 0$ and concave down for $x > 0$
     (d)   concave down for $x < 0$ and concave up for $x > 0$
     (e)   never concave up nor concave down

63. The graph of $y = e^{-1/x^2}$, $|x| > 2$, is

     (a)   concave up
     (b)   concave down
     (c)   concave up for $x < 0$ and concave down for $x > 0$
     (d)   concave down for $x < 0$ and concave up for $x > 0$
     (e)   never concave up nor concave down

64. The derivative $\dfrac{d}{dx} \log_3(\cos x)$ equals

     (a) $\dfrac{\sin x \cos x}{\ln 3}$

     (b) $-\dfrac{\ln 3 \cdot \sin x}{\cos x}$

(c) $\quad -\dfrac{\cos x}{\ln 3 \cdot \sin x}$

(d) $\quad -\dfrac{\sin x}{\ln 3 \cdot \cos x}$

(e) $\quad -\dfrac{\ln 3 \cdot \cos x}{\sin x}$

65. The derivative $\dfrac{d}{dx} 3^{x \ln x}$ equals

   (a) $\quad \ln 3 \cdot [x \ln x]$

   (b) $\quad (x \ln x) \cdot 3^{x \ln x - 1}$

   (c) $\quad 3^{x \ln x}$

   (d) $\quad \ln 3 \cdot [1 + \ln x]$

   (e) $\quad \ln 3 \cdot [1 + \ln x] \cdot 3^{x \cdot \ln x}$

66. The value of the limit $\lim_{h \to 0}(1 + h^2)^{1/h^2}$ is

   (a) $\quad e$

   (b) $\quad e - 1$

   (c) $\quad 1/e$

   (d) $\quad e^2$

   (e) $\quad 1$

67. Using logarithmic differentiation, the value of the derivative $\dfrac{x^2 \ln x}{e^x}$ is

   (a) $\quad \dfrac{\ln x}{e^x}$

   (b) $\quad \dfrac{x^2}{\ln x}$

   (c) $\quad \dfrac{x^2 \ln x}{e^x}$

   (d) $\quad \left(\dfrac{2}{x} + \dfrac{1}{x \ln x} - 1\right) \cdot \dfrac{x^2 \ln x}{e^x}$

   (e) $\quad \left(\dfrac{2}{x} - \dfrac{1}{x \ln x} - 1\right)$

68. The derivative of $f(x) = \mathrm{Sin}^{-1}(x \cdot \ln x)$ is

   (a) $\quad \dfrac{1 + \ln x}{x^2 \cdot \ln^2 x}$

   (b) $\quad \dfrac{1}{\sqrt{1 - x^2 \cdot \ln^2 x}}$

(c) $\dfrac{\ln x}{\sqrt{1-x^2}\cdot \ln^2 x}$

(d) $\dfrac{1+\ln x}{\sqrt{1-x^2}\cdot \ln^2 x}$

(e) $\dfrac{1+\ln x}{\sqrt{1-x^2}}$

69. The value of the derivative of $\operatorname{Tan}^{-1}(e^x \cdot \cos x)$ is

(a) $\dfrac{e^x}{1+e^{2x}\cos^2 x}$

(b) $\dfrac{e^x \sin x}{1+e^{2x}\cos^2 x}$

(c) $\dfrac{e^x \cos x}{1+e^{2x}\cos^2 x}$

(d) $\dfrac{e^x(\cos x - \sin x)}{1+\cos^2 x}$

(e) $\dfrac{e^x(\cos x - \sin x)}{1+e^{2x}\cos^2 x}$

70. The value of the integral $\int \log_3 x \, dx$ is

(a) $x \ln x - \dfrac{x}{\ln 3} + C$

(b) $x \log_3 x - x + C$

(c) $x \log_3 x - \dfrac{x}{\ln 3} + C$

(d) $x \log_3 x - \dfrac{x}{3} + C$

(e) $x \ln x - \dfrac{x}{3} + C$

71. The value of the integral $\int_0^1 5^{x^2} \cdot x \, dx$ is

(a) $\dfrac{4}{\ln 5}$

(b) $\dfrac{\ln 5}{2}$

(c) $\dfrac{\ln 5}{\ln 2}$

(d) $\dfrac{2}{\ln 5}$

(e) $\dfrac{\ln 2}{\ln 5}$

72. The value of the integral $\int x \cdot 2^x \, dx$ is

(a) $\dfrac{x \cdot 2^x}{\ln 2} - \dfrac{2^x}{\ln^2 2} + C$

(b) $\dfrac{x \cdot 2^x}{\ln 2} + \dfrac{2^x}{\ln^2 2} + C$

(c) $\dfrac{x \cdot 2^x}{\ln 2} - \dfrac{2^x}{\ln 2} + C$

(d) $\dfrac{2^x}{\ln 2} - 2^x + C$

(e) $\dfrac{x}{\ln 2} - \dfrac{2^x}{\ln 2} + C$

73. A petri dish contains 7,000 bacteria at 10:00 a.m. and 10,000 bacteria at 1:00 p.m. How many bacteria will there be at 4:00 p.m.?

(a) 700000
(b) 10000
(c) 100000
(d) 10000/7
(e) 100000/7

74. There are 5 grams of a radioactive substance present at noon on January 1, 2005. At noon on January 1 of 2009 there are 3 grams present. When will there be just 2 grams present?

(a) $t = 5.127$, or in early February of 2010
(b) $t = 7.712$, or in mid-August of 2012
(c) $t = 7.175$, or in early March of 2012
(d) $t = 6.135$, or in early February of 2011
(e) $t = 6.712$, or in mid-August of 2011

75. If \$8000 is placed in a savings account with 6% interest compounded continuously, then how large is the account after ten years?

(a) 13331.46
(b) 11067.35
(c) 14771.05
(d) 13220.12
(e) 14576.95

76. A wealthy uncle wishes to fix an endowment for his favorite nephew. He wants the fund to pay the young fellow $1,000,000 in cash on the day of his thirtieth birthday. The endowment is set up on the day of the nephew's birth and is locked in at 8% interest compounded continuously. How much principle should be put into the account to yield the necessary payoff?

    (a) 88,553.04
    (b) 90,717.95
    (c) 92,769.23
    (d) 91,445.12
    (e) 90,551.98

77. The values of $\text{Sin}^{-1}1/2$ and $\text{Tan}^{-1}\sqrt{3}$ are

    (a) $\pi/4$ and $\pi/3$
    (b) $\pi/3$ and $\pi/2$
    (c) $\pi/2$ and $\pi/3$
    (d) $\pi/6$ and $\pi/3$
    (e) $\pi/3$ and $\pi/6$

78. The value of the integral $\int \dfrac{dx}{4+x^2}\, dx$ is

    (a) $\dfrac{1}{2}\text{Tan}^{-1}\left(\dfrac{x}{2}\right)+C$

    (b) $\dfrac{1}{2}\text{Tan}^{-1}\left(\dfrac{x}{4}\right)+C$

    (c) $\dfrac{1}{4}\text{Tan}^{-1}\left(\dfrac{x}{2}\right)+C$

    (d) $\dfrac{1}{2}\text{Tan}^{-1}\left(\dfrac{x^2}{2}\right)+C$

    (e) $\dfrac{1}{2}\text{Tan}^{-1}\left(\dfrac{2}{x}\right)+C$

79. The value of the integral $\int \dfrac{e^x\, dx}{\sqrt{1-e^{2x}}}\, dx$

    (a) $\text{Cos}^{-1}e^x+C$
    (b) $\text{Sin}^{-1}e^{2x}+C$
    (c) $\text{Sin}^{-1}e^{-x}+C$
    (d) $\text{Cos}^{-1}e^{-2x}+C$
    (e) $\text{Sin}^{-1}e^x+C$

80. The value of the integral $\int_1^{\sqrt{2}} \dfrac{x\,dx}{x^2\sqrt{x^4-1}}\,dx$ is

   (a) $\dfrac{\pi}{3}$

   (b) $\dfrac{-\pi}{4}$

   (c) $\dfrac{-\pi}{6}$

   (d) $\dfrac{\pi}{4}$

   (e) $\dfrac{\pi}{6}$

81. The value of the integral $\int x^2 \ln x \, dx$ is

   (a) $\dfrac{x^3}{3} \ln x - \dfrac{x^3}{9} + C$

   (b) $\dfrac{x^2}{3} \ln x - \dfrac{x^2}{9} + C$

   (c) $\dfrac{x^3}{2} \ln x - \dfrac{x^3}{6} + C$

   (d) $\dfrac{x^3}{9} \ln x - \dfrac{x^3}{3} + C$

   (e) $\dfrac{x^5}{3} \ln x - \dfrac{x^3}{6} + C$

82. The value of the integral $\int_0^1 e^x \sin x \, dx$ is

   (a) $e \cdot \cos 1 - e \cdot \sin 1 + 1$
   (b) $e \cdot \sin 1 - e \cdot \cos 1 - 1$
   (c) $e \cdot \sin 1 - e \cdot \cos 1 + 1$
   (d) $e \cdot \sin 2 - e \cdot \cos 2 + 1$
   (e) $e \cdot \sin 2 + e \cdot \cos 2 - 1$

83. The value of the integral $\int x \cdot e^{2x} \, dx$ is

   (a) $\dfrac{xe^x}{2} - \dfrac{e^x}{4} + C$

   (b) $\dfrac{xe^{2x}}{4} - \dfrac{e^{2x}}{2} + C$

   (c) $\dfrac{xe^x}{4} - \dfrac{e^x}{2} + C$

(d) $\dfrac{xe^{2x}}{2} - \dfrac{e^x}{4} + C$

(e) $\dfrac{xe^{2x}}{2} - \dfrac{e^{2x}}{4} + C$

84. The value of the integral $\displaystyle\int \dfrac{dx}{x(x+1)}$ is

(a) $\ln|x+1| - \ln|x| + C$
(b) $\ln|x-1| - \ln|x+1| + C$
(c) $\ln|x| - \ln|x+1| + C$
(d) $\ln|x| - \ln|x| + C$
(e) $\ln|x+2| - \ln|x+1| + C$

85. The value of the integral $\displaystyle\int \dfrac{dx}{x(x^2+4)}$ is

(a) $\dfrac{1}{2}\ln|x| - \dfrac{1}{8}\ln(x^2+2) + C$

(b) $\dfrac{1}{4}\ln|x| - \dfrac{1}{8}\ln(x^2+4) + C$

(c) $\dfrac{1}{8}\ln|x| - \dfrac{1}{4}\ln(x^2+4) + C$

(d) $\dfrac{1}{2}\ln|x| - \dfrac{1}{8}\ln(x^2+1) + C$

(e) $\dfrac{1}{8}\ln|x| - \dfrac{1}{2}\ln(x^2+4) + C$

86. The value of the integral $\displaystyle\int \dfrac{dx}{(x-1)^2(x+1)}$ is

(a) $\dfrac{1}{2}\ln|x-1| + \dfrac{1}{x-1} - \dfrac{1}{2}\ln|x+1| + C$

(b) $\dfrac{-1}{2}\ln|x-1| - \dfrac{1}{(x-1)^2} + \dfrac{1}{2}\ln|x+1| + C$

(c) $\dfrac{-1}{2}\ln|x-1| + \dfrac{1}{x-1} + \dfrac{1}{4}\ln|x+1| + C$

(d) $\dfrac{-1}{2}\ln|x-1| - \dfrac{1}{x-1} + \dfrac{1}{2}\ln|x+1| + C$

(e) $\dfrac{-1}{4}\ln|x-1| - \dfrac{1/2}{x-1} + \dfrac{1}{4}\ln|x+1| + C$

87. The value of the integral $\int_1^2 x\sqrt{1+x^2}\,dx$ is

(a) $\frac{1}{4}5^{3/2} - \frac{1}{4}2^{3/2}$

(b) $\frac{1}{3}4^{3/2} - \frac{1}{3}3^{3/2}$

(c) $\frac{1}{3}5^{3/2} - \frac{1}{3}2^{3/2}$

(d) $\frac{1}{5}4^{3/2} - \frac{1}{5}3^{3/2}$

(e) $\frac{1}{3}2^{3/2} - \frac{1}{3}5^{3/2}$

88. The value of the integral $\int_0^{\pi/4} \dfrac{\sin x \cos x}{1 + \cos^2 x}\, dx$

(a) $\ln \dfrac{\sqrt{3}}{5}$

(b) $\ln \dfrac{\sqrt{5}}{2}$

(c) $\dfrac{1}{2} \ln \dfrac{\sqrt{3}}{2}$

(d) $\ln \dfrac{\sqrt{2}}{3}$

(e) $\ln \dfrac{2}{\sqrt{3}}$

89. The value of the integral $\int_0^1 e^x \sin(1 + e^x)\, dx$ is

(a) $-\sin(1 + e) + \cos 2$

(b) $-\sin(1 - e) + \sin 2$

(c) $-\cos(1 + e) + \cos 2$

(d) $\cos(1 + e) - \cos 2$

(e) $\cos(1 + e) + \cos 2$

90. The value of the integral $\int_0^\pi \sin^4 x\, dx$ is

(a) $\dfrac{5\pi}{8}$

(b) $\dfrac{3\pi}{8}$

(c) $\dfrac{5\pi}{6}$

(d) $\dfrac{3\pi}{10}$

(e) $\dfrac{2\pi}{5}$

91. The value of the integral $\int_0^\pi \sin^2 x \cos^2 x \, dx$ is

(a) $\dfrac{\pi}{6}$

(b) $\dfrac{\pi}{4}$

(c) $\dfrac{\pi}{3}$

(d) $\dfrac{\pi}{2}$

(e) $\dfrac{\pi}{8}$

92. The value of the integral $\int_0^{\pi/4} \tan^2 x \, dx$ is

(a) $1 - \dfrac{\pi}{3}$

(b) $2 - \dfrac{\pi}{4}$

(c) $1 - \dfrac{\pi}{2}$

(d) $1 - \dfrac{\pi}{4}$

(e) $4 - \pi$

93. A solid has base in the x-y plane that is the circle of radius 1 and center the origin. The vertical slice parallel to the y-axis is a semi-circle. What is the volume?

(a) $\dfrac{4\pi}{3}$

(b) $\dfrac{2\pi}{3}$

(c) $\dfrac{\pi}{3}$

(d) $\dfrac{8\pi}{3}$

(e) $\dfrac{\pi}{6}$

94. A solid has base in the x-y plane that is a square with center the origin and vertices on the axes. The vertical slice parallel to the y-axis is an equilateral triangle. What is the volume?

(a) $\dfrac{2\sqrt{3}}{3}$

(b) $\dfrac{\sqrt{3}}{3}$

(c) $\sqrt{3}$

(d) $\sqrt{3} + 3$

(e) $3\sqrt{3}$

95. The planar region bounded by $y = x^2$ and $y = x$ is rotated about the line $y = -1$. What volume results?

(a) $\dfrac{11\pi}{15}$

(b) $\dfrac{7\pi}{15}$

(c) $\dfrac{7\pi}{19}$

(d) $\dfrac{8\pi}{15}$

(e) $\dfrac{2\pi}{15}$

96. The planar region bounded by $y = x$ and $y = \sqrt{x}$ is rotated about the line $x = -2$. What volume results?

(a) $\dfrac{4\pi}{5}$

(b) $\dfrac{4\pi}{7}$

(c) $\dfrac{9\pi}{5}$

(d) $\dfrac{4\pi}{3}$

(e) $\dfrac{11\pi}{5}$

97. A bird is flying upward with a leaking bag of seaweed. The sack initially weights 10 pounds. The bag loses 1/10 pound of liquid per minute, and the bird increases its altitude by 100 feet per minute. How much work does the bird perform in the first six minutes?

(a) 5660 foot-pounds
(b) 5500 foot-pounds

    (c)   5800 foot-pounds
    (d)   5820 foot-pounds
    (e)   5810 foot-pounds

98. The average value of the function $f(x) = \sin x - x$ on the interval $[0, \pi]$ is

    (a) $\dfrac{3}{\pi} - \dfrac{\pi}{4}$

    (b) $\dfrac{2}{\pi} - \dfrac{\pi}{3}$

    (c) $\dfrac{2}{\pi} - \dfrac{\pi}{2}$

    (d) $\dfrac{4}{\pi} - \dfrac{\pi}{4}$

    (e) $\dfrac{1}{\pi} - \dfrac{\pi}{2}$

99. The integral that equals the arc length of the curve $y = x^3$, $1 \le x \le 4$, is

    (a) $\displaystyle\int_1^4 \sqrt{1 + x^4}\, dx$

    (b) $\displaystyle\int_1^4 \sqrt{1 + 9x^2}\, dx$

    (c) $\displaystyle\int_1^4 \sqrt{1 + x^6}\, dx$

    (d) $\displaystyle\int_1^4 \sqrt{1 + 4x^4}\, dx$

    (e) $\displaystyle\int_1^4 \sqrt{1 + 9x^4}\, dx$

100. The Simpson's Rule approximation to the integral $\displaystyle\int_0^1 \dfrac{dx}{\sqrt{1 + x^2}}\, dx$ with $k = 4$ is

    (a) ≈ 0.881
    (b) ≈ 0.895
    (c) ≈ 0.83
    (d) ≈ 0.75
    (e) ≈ 0.87

SOLUTIONS

1. (a),	2. (c),	3. (b),	4. (e),	5. (e),	6. (d),	7. (b),
8. (a),	9. (c),	10. (d),	11. (e),	12. (b),	13. (c),	14. (d),
15. (e),	16. (a),	17. (c),	18. (d),	19. (c),	20. (e),	21. (a),
22. (d),	23. (b),	24. (c),	25. (c),	26. (a),	27. (d),	28. (e),
29. (c),	30. (b),	31. (e),	32. (e),	33. (c),	34. (c),	35. (a),
36. (a),	37. (d),	38. (e),	39. (b),	40. (d),	41. (e),	42. (b),
43. (a),	44. (b),	45. (c),	46. (d),	47. (c),	48. (d),	49. (b),
50. (c),	51. (b),	52. (a),	53. (d),	54. (d),	55. (a),	56. (c),
57. (b),	58. (c),	59. (e),	60. (e),	61. (d),	62. (a),	63. (a),
64. (d),	65. (e),	66. (a),	67. (d),	68. (d),	69. (e),	70. (c),
71. (d),	72. (a),	73. (e),	74. (c),	75. (e),	76. (b),	77. (d),
78. (a),	79. (e),	80. (e),	81. (a),	82. (c),	83. (e),	84. (c),
85. (b),	86. (e),	87. (c),	88. (e),	89. (c),	90. (b),	91. (e),
92. (d),	93. (b),	94. (a),	95. (b),	96. (a),	97. (d),	98. (c),
99. (e),	100. (a)					

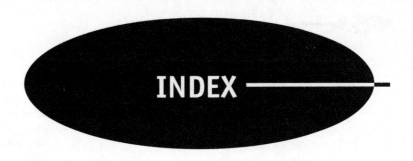
INDEX

Index

Index

Index

ABOUT THE AUTHOR

Steven G. Krantz is the Chairman of the Mathematics Department at Washington University in St. Louis. An award-winning teacher and author, Dr. Krantz has written more than 30 books on mathematics, including a best-seller.